Differential Equations with *Mathematica*™

Second Edition

Revised for Mathematica 3.0

Kevin R. Coombes
Brian R. Hunt
Ronald L. Lipsman
John E. Osborn
Garrett J. Stuck

All of the University of Maryland at College Park

John Wiley & Sons, Inc.
New York • Chichester • Weinheim • Brisbane • Singapore • Toronto

ISBN 0-471-17696-6

Printed in the United States of America

10   9   8   7   6   5   4   3   2   1

Printed and bound by Malloy Lithographing, Inc.

#39485342

Preface

> *As the subject matter of differential equations continues to grow, as new technologies become commonplace, as old areas of application are expanded, and as new ones appear on the horizon, the content and viewpoint of courses and their textbooks must also evolve.*

Boyce & DiPrima, **Elementary Differential Equations**, Sixth Edition.

Traditional introductory courses in ordinary differential equations (ODE) have concentrated on teaching a repertoire of techniques for finding formula solutions of various classes of differential equations. Typically, the result was rote application of formula techniques without a serious qualitative understanding of such fundamental aspects of the subject as stability, asymptotics, dependence on parameters, and numerical methods. These fundamental ideas are difficult to teach because they have a great deal of geometrical content and, especially in the case of numerical methods, involve a great deal of computation. Modern mathematical software systems, which are particularly effective for geometrical and numerical analysis, can help to overcome these difficulties. This book changes the emphasis in the traditional ODE course by using a mathematical software system to introduce numerical methods, geometric interpretation, symbolic computation, and qualitative analysis into the course in a basic way.

The mathematical software system we use is *Mathematica*. (This book is also available in a *Maple* version.) We assume that the user has no prior experience with *Mathematica*. We include concise instructions for using *Mathematica* on three popular computer platforms: *Windows*, *Macintosh*, and the *X Window System*. This book is not a comprehensive introduction or reference manual to either *Mathematica* or any of the computer platforms. Instead, it focuses on the specific features of *Mathematica* that are useful for analyzing differential equations. It also describes the features of the *Mathematica* "Notebook" interface that are necessary for creating a finished document.

This supplement can easily be used in conjunction with most ODE texts. It addresses the standard topics in ODE, but with a substantially different emphasis.

We had two basic goals in mind when we introduced this supplement into our course. First, we wanted to deepen students' understanding of differential equations by giving them a new tool, a mathematical software system, for analyzing differential equations. Second, we wanted to bring students to a level of expertise in the mathematical software system that would allow them to use it in other mathematics, engineering, or science courses. We believe that we have achieved these goals in our own classes. We hope this supplement will be useful to students and instructors on other campuses in achieving the same goals.

Acknowledgements

We thank Peter Olver, Jonathan Rosenberg, Larry Shampine, Shagi-Di Shih, and Nancy Stanton, all of whom contributed to this book. We are grateful to the 40 or so of our colleagues who have used the first edition and other versions of this material in their classes at the University of Maryland at College Park. We are particularly indebted to the many students who have used these materials, and have communicated to us their comments and suggestions. Finally, we thank Barbara Holland, mathematics editor at Wiley, for her enthusiastic and thoughtful support of our project.

Kevin Coombes
Brian Hunt
Ron Lipsman
John Osborn
Garrett Stuck

College Park, Maryland
August 1, 1997

Contents

Chapter 1

Introduction

We begin by describing the philosophy behind our approach to the study of ordinary differential equations. This philosophy has its roots in the way we understand, use, and apply differential equations; it has influenced our teaching and guided the development of this book. This chapter also contains two user's guides, one for students and one for instructors.

Guiding Philosophy

In scientific inquiry, when we are interested in understanding, describing, or predicting some complex phenomenon, we use the technique of *mathematical modeling*. In this approach, we describe the state of a physical, biological, economic, or other system by one or more functions of one or several variables. For example, the position $s = s(t)$ of a particle is a function of the time t; the temperature $T = T(x, y, z, t)$ in a body is a function of the position (x, y, z) and the time t; the gravitational or electromagnetic force on an object is a function of its position; the money supply is a function of time; the populations $x = x(t)$ and $y = y(t)$ of two competing species are functions of time. Next, we attempt to formulate, in mathematical terms, a fundamental law governing the phenomenon. Typically, this formulation results in one or more differential equations; *i.e.*, equations involving derivatives of the functions describing the state of the system with respect to the variables they depend on. Frequently, the functions depend on only one variable, and the differential equation is called *ordinary*. To be specific, if $x = x(t)$ denotes the function describing the state of our system, an *ordinary differential equation* in $x(t)$ might involve $x'(t)$, $x''(t)$, higher derivatives, or other known functions of t. By contrast, in a *partial differential equation*, the functions depend on several variables.

   In addition to the fundamental law, we usually describe the initial state of the system. We express this state mathematically by specifying *initial values* $x(0)$, $x'(0)$, *etc.* In this way, we arrive at an *initial value problem*—an ordinary differential equation together with initial conditions. If we can solve the initial value problem, then the solution is a function $x(t)$ that predicts the future state of the system. Using the solution, we can describe qualitative or quantitative properties of the system. At this stage, we compare the values predicted by $x(t)$ against experimental data that we accumulate by observing the system. If the experimental data and the function values match, we congratulate ourselves on a job well done and go on to the next challenge. If they do not match, we go back, refine the model, and start again. Even if we are satisfied with the model, we may later arrive at a time when new technology, new requirements, or newly discovered features of the system render our old data obsolete. Again, we respond by reexamining the model.

The subject of differential equations consists in large part of building and solving mathematical models. Many results and methods have been developed for this purpose. These results and methods fall within one or more of the following themes:

(1) Existence and uniqueness of solutions
(2) Dependence of solutions on initial values
(3) Derivation of formulas for solutions
(4) Numerical calculation of solutions
(5) Graphical analysis of solutions
(6) Qualitative analysis of differential equations and their solutions.

The basic results on existence and uniqueness (theme 1), and dependence on initial values (theme 2), form the foundation of the subject of differential equations. The derivation of formula solutions (theme 3) is a rich and important part of the subject; a variety of methods have been developed for finding formula solutions to special classes of equations. Although many equations can be solved exactly, many others cannot. However, any equation, solvable or not, can be analyzed using numerical methods (theme 4), graphical methods (theme 5), and qualitative methods (theme 6). The ability to obtain qualitative and quantitative information without the aid of an explicit formula solution is crucial. That information may suffice to analyze and describe the original phenomenon (which led to the model, which gave rise to the differential equation).

Traditionally, introductory courses in differential equations focused on methods for deriving exact solutions to special types of equations, and included some simple numerical and qualitative methods. The human limitations involved in compiling numerical or graphical data were formidable obstacles to implementing more advanced qualitative or quantitative methods. Computer platforms have reduced these obstacles. Sophisticated software and mainframe computers enhanced the use of quantitative and qualitative methods in the theory and applications of differential equations. With the arrival of comprehensive mathematical software systems on personal computers, this modern approach has become accessible to undergraduates.

In this book, we use the mathematical software system *Mathematica* to implement this new approach. We introduce symbolic, numerical, graphical, and qualitative techniques, and show how to use *Mathematica* to analyze differential equations and their solutions with these techniques. As you encounter each new technique, you should ask: Under which of the six themes listed above does it fit? Answering that question will help you to organize the course material.

In order to take advantage of this modern approach, you must learn to use a computer platform effectively. We give detailed instructions for using *Mathematica* on three specific types of computers. In this way, we minimize the time required to learn to use the computer platform.

Finally, the software system we have chosen provides an additional valuable opportunity: it enables the production of "Notebooks" that combine textual commentary with the numerical, symbolic, and graphical output of the mathematical software system. Engineers and scientists have to develop not only skills in analyzing problems and interpreting solutions, but also the ability to present coherent conclusions in a logical and convincing style. You should use the Notebook capabilities of the software to submit solutions to the computer assignments in such a style. This is excellent preparation for the professional requirements that lie ahead.

Student's Guide

The chapters of this book can be divided into three classes: descriptions of the computer platform, supplementary material on ordinary differential equations (ODE), and computer problem sets. Here is a brief description of the contents.

The computer platform chapters require no prior knowledge of computers. Chapter 2 explains how to start *Mathematica* in *Windows 95*, on the *Macintosh*, and in the *X Window System*. Chapter 3 introduces some basic *Mathematica* commands. You should work through Chapters 2 and 3 while sitting at your computer. Then you should work the problems in Problem Set A. These steps will bring you to a minimal level of competence in the use of *Mathematica*. You should complete these tasks within one week after the semester begins. After that, read Chapter 4, which contains detailed instructions for manipulating the *Mathematica* Notebook interface on your computer. Finally, Chapter 8 is an introduction to the more sophisticated *Mathematica* commands that will be useful in later problem sets. It is best to read this chapter after working some problems in Problem Set B, but before starting on Problem Set C.

Keep in mind that the computer platform chapters are intended to teach you enough about *Mathematica* to study ordinary differential equations. You can explore *Mathematica* in more depth using its extensive online help resources. You can also consult any of a number of books about *Mathematica*. The definitive reference is: Stephen Wolfram, **The Mathematica Book**, Third Edition, Cambridge University Press, 1996. An elementary introduction is: **The Mathematica Primer**, Cambridge University Press, 1997, by the authors of this book.

The eight ODE chapters (5–7, 9–13) are intended to supplement the material in your text. The emphasis in this book differs from that found in a traditional ODE text. The main difference is less emphasis on the search for exact formula solutions, and greater emphasis on qualitative, graphical, and numerical analysis of the equations and their solutions. Furthermore, the commands for analyzing differential equations with *Mathematica* appear in these chapters.

The third collection of documents in this book consists of six sets of computer problems. The problem sets form an integral part of the course. Solving these problems will expose you to the qualitative, graphical, and numerical features of the course. Each set contains about fifteen problems; your instructor will decide which problems to assign, and when and how you are to submit them.

You can most profitably attack the problem sets if you plan to do them in two distinct sessions. Begin by reading the problems and thinking about the issues involved. Jot down some notes about how *Mathematica* can help you solve the problems. Although it is tempting to sit at the computer and start typing immediately, you will find that approach to be inefficient and potentially frustrating. Some preparation with pencil and paper will be useful. Then go to the computer and start solving the problems. If you get stuck, save your work and go on to the next problem. You should expect to spend at least one hour per problem. If there are things you'd like to discuss with your instructor, print out the relevant parts of the Notebook to take with you. Then log out, go home, and try to digest what you've done—decide what's right, what's wrong, what could be done better, what surprises you encountered, *etc.* Talk to your instructor or your peers about anything you don't understand. This first session should be attempted a week before the assignment is due. After you have reviewed your output and obtained answers to your questions, you are ready for your second session. At this point, you should fill the gaps, correct your mistakes, and polish your Notebook. This will take another hour or so per problem. Although you may find yourself spending extra time on the first few problems, if you read the *Mathematica* chapters carefully, and follow the suggestions above, you should steadily increase your level of competence in using *Mathematica*.

The end of this book contains two useful sections: a Glossary and a collection of Sample Notebook Solutions. The Glossary contains a brief summary of some *Mathematica* commands, options, and built-in functions. The Sample Notebook Solutions show how we solved several problems from this book. These samples will serve as guides when you prepare your own Notebook solutions. Emulate them. Strive to prepare coherent, organized solutions. Combine *Mathematica*'s input, output, and graphics with your own textual commentary and analysis of the problem. Edit the final version of your solution to remove syntax errors and false starts. You will soon take pride in submitting complete, polished solutions to the problems.

Instructor's Guide

The philosophy that guided the writing of this book is explained at the beginning of this chapter. Here is a capsule summary of that philosophy. We seek:

- To guide students into a more interpretive mode of thinking.
- To use a mathematical software system to enhance students' ability to compute symbolic and numerical solutions, and to perform qualitative and graphical analysis of differential equations.

- To develop course material that reflects the current state of ODE and emphasizes the mathematical modeling of physical problems.
- To minimize the time required to learn to use the computer platform.

Our material consists of computer platform documents, ODE supplements, and computer problem sets. Here are our recommendations for integrating this material into the course.

Platform Documents

These are discussed in the preceding section. Proper use of these documents will quickly propel students to a reasonable level of proficiency. Students should read Chapters 2 and 3, and work Problem Set A within the first week of the semester. They should read Chapter 4 immediately after that, but they can delay reading Chapter 8 until after they start Problem Set B.

ODE Chapters

These eight chapters supplement the material in a traditional text. We use *Mathematica* to study differential equations using symbolic, numerical, graphical, and qualitative methods. We emphasize the following topics: stability, comparison methods, numerical methods, direction fields, and phase portraits. These topics are not emphasized to the same degree in a traditional text. We suggest that you incorporate this new emphasis into your class discussions. You should devote some class time to each chapter. Specific guidelines are difficult to prescribe, and the required time varies with each chapter, but on average we spend 30–40 minutes per chapter in class discussion. Of course, students will pay greater attention to the chapters if the material appears on quizzes and exams.

The structure of this book requires that numerical methods be discussed early in the course, immediately after the discussion of first order equations. The discussion of numerical methods should be directed toward the use of *Mathematica*'s numerical ODE solver.

Computer Problem Sets

There are six computer problem sets. The topics addressed in the problem sets are:

(A) Practice with *Mathematica*
(B) First Order Equations
(C) Numerical Solutions of Differential Equations
(D) Second Order Equations
(E) Series Solutions and Laplace Transforms
(F) Systems of Equations

Problem Set A is a practice set designed to acquaint students with the basic symbolic and graphical capabilities of *Mathematica*, and to reacclimate them to calculus. It should be assigned on the first day of class, to be completed within one week. It is best to assign each remaining problem set at least two weeks before the due date. The course material corresponding to the problem set should be covered during those two weeks. It is good practice to set aside a small amount of class time to discuss questions that arise in students' first attempts to work the problems. That should typically be 3–5 days before the due date. (In the student guide, we suggested a two-stage plan of attack for completing the problem sets. Your class discussion should ideally occur after the first stage.)

We recommend that you assign all of Problem Set A and 3–6 problems from the other problem sets. Students should be required not only to analyze the problems critically, but also to present their analysis in coherent English and mathematics, displayed appropriately on their printouts. To accomplish this, students must master the editing features in the Notebook interface of *Mathematica*. These features are discussed in the computer platform chapters. This skill is very important: Engineers and scientists do not just solve problems; they must also present their ideas in cogent and convincing fashion. We expect students to do the same in this course.

It is important to keep in mind that computer files can be copied effortlessly. You should choose a policy regarding student collaboration, and make this policy clear. For example, you might allow students to discuss the assignments, but require them to create the solution Notebooks themselves. Or you might allow students to work jointly and to submit collective solutions. We have found that allowing students to collaborate (in teams of no more than three) can be very beneficial. You should also vary the problems you assign from semester to semester.

Hardware and Software Considerations

Before using these materials, you should make sure that your institution has enough machines with enough computing power. *Mathematica* requires a computer with at least 16 megabytes of random access memory (RAM).

New versions of software appear frequently. When a complex system like *Mathematica* changes, many commands work better than they did before, some work differently, and a few may no longer work at all. As this book goes to press, the current version is *Mathematica 3.0*; this is the version we describe. Most problems in the book can also be solved using version 2.1 or 2.2.

Generalities

You do not have to be a *Mathematica* expert to use this course supplement. On the other hand, both you and your students will benefit if you spend some time working with *Mathematica*. We recommend that, in addition to working through Problem Set A, you work through the problems you assign in each set, preferably on the same computers that the students will use.

Our practice is to spend little class time discussing *Mathematica* per se. We try not to lose sight of the fact that we are teaching a course in *mathematics*, not *Mathematica*. We expect students to gain expertise in *Mathematica* from this book, from experimentation, from *Mathematica*'s extensive online help, from talking to each other, from questions addressed to the instructor after class and in office hours, and by consulting other *Mathematica* references.

Students often try to establish the following mode in their mathematics courses: "Give me a formula, I will give you a number!" However, we want students to understand the motives and methods behind a formula and its use. In the graphical environment offered by *Mathematica*, students sometimes try to establish a new mode: "Give me a program, I will give you a picture!" We ask not only that students produce a picture, but also that they interpret it, analyze the phenomenon it represents, and describe clearly and coherently the conclusions they derive from it.

Sample Syllabus

In this section, we describe how this supplement might be used with two different elementary differential equations texts: **Elementary Differential Equations,** Sixth Edition, by Boyce & DiPrima (Wiley, 1996) and **Differential Equations and Boundary Value Problems,** by Edwards and Penney (Prentice-Hall, 1996).

This supplement contains enough material to address the topics in the first nine chapters of Boyce & DiPrima or the first eight chapters of Edwards & Penney. This material is more than ample for one semester or two quarters of study. If the course is given in a 15-week semester, then it is likely that portions of these chapters would have to be omitted. For example, at the University of Maryland, College Park, we omit Chapters 5 and 6 of Boyce & DiPrima. Because of the emphasis in this supplement on *Mathematica*'s numerical ODE solver, it is imperative that numerical methods be treated early in the course.

The following schedules are for a typical semester. We assume a MWF schedule with 38 lectures and three days for exams.

Syllabus for use with Boyce & DiPrima

B & D Sections	Material from this book Chapters	Prob. Sets	No. of Lectures
2.1–2.8	1–6	A, B	10
8.1–8.4	7, 8	C	3
3.1–3.9	9	D	10
4.1–4.2	12		1
7.1–7.7	12	F	7
9.1–9.5	13	F	7

Syllabus for use with Edwards & Penney

E & P Sections	Material from this book Chapters	Prob. Sets	No. of Lectures
1.1–1.5	1–6	A, B	7
2.1–2.6	7, 8	C	6
3.1–3.6	9	D	8
4.1–4.3	12	F	4

| 5.1–5.3, 6.1–6.5 | 13 | F | 8 |
| 7.1–7.6 | 11 | E | 5 |

The following schedule indicates where exams and problem set due dates could be placed for the Boyce & DiPrima syllabus. Each problem set should be assigned at least two weeks in advance of the due date.

Item	Lecture Number
Problem Set A	5
Exam 1	11
Problem Set B	13
Problem Set C	20
Exam 2	25
Problem Set D	27
Exam 3	38
Problem Set F	40

Instructors have a great deal of flexibility in deciding how to integrate the material from this supplement into a course. By choosing among the ODE chapters (Chapters 5–7 and 9–13), the instructor can decide which aspect of the course to emphasize. In order to emphasize numerical methods, you could stress Chapter 7 and Problem Set C. In order to emphasize symbolic computation, you could stress Chapters 5, 10, and 11 and Problem Set E. In order to emphasize geometric methods, you could stress Chapters 6, 9, and 13 and Problem Sets B and D. Chapter 12 incorporates both symbolic and geometric methods.

By adjusting the number of problems assigned from each problem set, the instructor can adjust the level of intensity. For example, some of our colleagues spread the problems out in a more uniform way during a semester, instead of assigning several substantial problem sets. Although we prefer to assign a few substantial problem sets, the material accommodates either mode.

The authors maintain a Web site for this book at

http://www.math.umd.edu/schol/ODE

Chapter 2

Getting Started
with *Mathematica*

In this chapter, we describe the basic skills needed to use *Mathematica* on several different computer platforms. In particular, we describe how to open, save, and print files. These are all features of the *interface* between *Mathematica* and the particular computer operating system that you are using. These instructions are distinct from the actual commands you will use to do mathematics with *Mathematica*. The next chapter introduces the mathematical aspects of *Mathematica*.

This chapter has three parts, one for each of the following user interfaces: *Windows 95*, *Macintosh*, and the *X Window System* (commonly found on *UNIX* workstations). You should find out what kind of computer and operating system you will be using, and then sit down at the computer with this guide and read the appropriate section. If you are using *Windows 95* or the *Macintosh* operating system on a networked computer, some of the features we describe below may be different for your particular installation. You may need to consult a local expert if you are using a different operating system (such as *NeXTStep*, *OS/2*, *Windows 3.1*, or *Windows NT*).

All of the systems we describe are window-based interfaces, which are manipulated using the keyboard and the mouse. Moving the mouse moves a small arrow (called a *pointer*) on the screen. To *click* on an object, place the pointer over the object and press the mouse button (if your mouse has more than one button, use the left button). To *double-click*, position the pointer and press the button twice in rapid succession. To *drag* an object, position the pointer over it, press and hold the mouse button, slide the object to where you want it, and then release the mouse button. A window can be dragged by its titlebar.

Mathematica in *Windows 95*

When you sit down at the computer, it may be turned off or it may be running a screensaver program. If it's off, turn it on and wait until it finishes booting. If a screensaver is running, move the computer's mouse to refresh the display. If all goes well, the screen should now contain various windows and/or icons.

In *Windows 95*, you manage programs, or *applications*, from the "Taskbar". The Taskbar usually appears as a slender gray rectangle across the bottom of the screen.

At its left end, there is a button labeled "Start". Pressing the "Start" button brings up a menu of choices, including Help, Settings, and Programs. Clicking on one of these choices either opens a submenu with more choices (submenus may be nested several layers deep) or launches an application.

In addition to the "Start" button, the Taskbar may contain other buttons. Each button represents an active application. When you start an application, a button is added to the Taskbar; clicking on the button will reactivate the application.

In addition to the Taskbar, the *Windows 95* desktop contains several icons. Double-clicking on the icon labeled "My Computer" will open a window containing icons that represent the disk drives and other hardware installed in the computer. Double-clicking on a drive icon opens a window showing the files and folders stored on the drive. Double-clicking on a file will start the application associated with the file, allowing you to view and modify the file's contents. Double-clicking on a folder will open a window displaying its files and subfolders.

If the computer is networked, then there will be an icon labeled "Network Neighborhood". Double-clicking on the "Network Neighborhood" icon will open a window showing the resources available on the network. Double-clicking on these icons may allow you to manipulate the files and folders stored on other computers on the network.

If a window is partially obscured, you can bring it to the front by clicking on any part of it, or by clicking on the corresponding button on the Taskbar. In the upper right-hand corner of a window are three small boxes. Clicking the box marked with an "X" will quit the application. Clicking the box marked with a rectangle will expand the window to fill the entire screen. Clicking the box marked with a small line will make the window disappear; you can make it reappear by clicking on the corresponding button on the Taskbar.

Online Help for *Windows 95*

If you have never used *Windows 95* before, it's probably a good idea to take the "Windows Tour". To do this, press the "Start" button, and then choose Help. This will open a help window. Click the tab labeled "Contents", then double-click on "Tour: Ten minutes to using Windows". The tour covers the basics of managing windows and applications.

Starting *Mathematica*

The *Mathematica* program uses a lot of memory, so you might want to close other applications before starting *Mathematica*. An application can be closed by choosing the Quit or Exit button in the File menu belonging to the application, or by clicking on the small box labeled with an "X" in the upper right-hand corner of the window.

Once you have closed the other applications, you are ready to start *Mathematica*. Press the "Start" button and choose Programs. The icon for the *Mathematica* application is a stellated icosahedron (a three-dimensional star) with the name

"Mathematica 3.0" next to it. (Make sure you get "Mathematica 3.0" and not "Mathematica 3.0 Kernel".) The location of this application will depend on your particular machine, so you may have to hunt around to locate it. Try looking in submenus that have the word "Application" or "Math" in their titles. When you've found the correct menu item, click on it. After a brief delay, three new windows will appear on the screen. The narrow window at the top is the *Mathematica menu bar*. The main blank window at the left is a *Notebook*, whose name is "Untitled-1". The window on the right is a *Palette*, which can be used to insert mathematical symbols into the Notebook. You can now start typing in the Notebook.

Saving Your Notebook on a Floppy Disk

To save your Notebook onto a disk, you'll need to buy some disks for the computer you are using (usually 3.5 inch, 1.44 megabyte High Density (HD) disks). These are inexpensive and available in many book stores and computer stores.

Insert a disk in the slot in the front of the computer. To find out what files are stored on the disk, move or hide windows until you can see the "My Computer" icon on the *Windows 95* desktop. Double-click on the "My Computer" icon. Move the mouse pointer to the icon representing the floppy drive. (The icon will be labeled "Floppy (A:)" or "Floppy (B:)". Make sure that you are pointing to the correct floppy drive.) Double-click on the floppy icon to view the files it contains.

If the disk is unformatted, or formatted for a different kind of computer, you will have to format it before you can use it. *Caution:* Formatting a disk removes all information from it. Follow the instructions above to make certain that you do not have important files on a disk before formatting it.

To format a disk, double-click the "My Computer" icon and point to the correct floppy drive. Press the right-hand mouse button to open a menu, and click on Format.... A dialog panel will appear. In this panel, you can give a name to the disk by typing in the "Label" box. To format an unformatted disk, choose "Full" as the format type. (A "Quick" format will only work on previously formatted disks.) Once you have selected the proper format type and typed in a name, click on the "Start" button in the dialog panel. A full format may take a few minutes.

Once you have a properly formatted disk inserted in the drive, make sure that your *Mathematica* window is active by clicking in it. A darkened titlebar indicates that the window is active. Now select Save As... from the File menu. A dialog panel will appear on the screen. Near the bottom of the panel is a box labeled "Drives:". The drive that has your disk will be either "A:" or "B:". Select the proper drive by clicking on its name. (If you don't see the name of the drive, use the arrow button to move around in the list of drives.) Once you have the proper drive selected, click in the box labeled "File Name:" (near the upper left-hand corner) and then type the name you'd like to use for the Notebook. The name should have the extension ".nb"; for example, "Problem1.nb". (Older versions of *Mathematica* saved files with a ".ma" extension.) Then click "OK". If all goes well, a copy of the file will be saved on the disk, and the name of the *Mathematica* window will change from "Untitled-1" to the name you typed.

Once a Notebook has a title, you can save further changes by simply selecting Save from the File menu. *Mathematica* automatically keeps track of the filename. If you want to save the Notebook under a different name, you'll have to use the Save As... button again. You should save your work frequently.

If the computer you are using is networked, and you have an account on the network, you may be able to save your Notebook on a central file server. Consult your system administrator for instructions.

Opening a Previously Saved Notebook

To open a saved Notebook, click on the File button, and then click on the Open button. A new dialog panel will appear. This panel is essentially identical to the one that appeared as a consequence of selecting the Save As... button. You can navigate in this panel by clicking on drives and/or directories until the file you want appears in the box labeled "File Name:". After selecting the file you want, click on the "OK" button. *Mathematica 3.0* Notebooks have the filename extension ".nb".

To open a new Notebook, click the New button in the File menu. You can open and save as many Notebooks as you like, using a different name for each one, as long as you don't use up the available disk space.

If you have several Notebooks open, you may want to close one or more. To close a Notebook, make it active by clicking in it or by selecting its name from the menu labeled Window, and then select Close from the File menu. *Mathematica* will prompt you to save the Notebook if the Notebook has changed since the last save. Do not attempt to close a Notebook by selecting Exit from the File menu; that will terminate your *Mathematica* session entirely.

Printing Your Notebook

To print a Notebook, first make sure the Notebook you want to print is the active window. Then select Print... from the File menu. A new panel will appear offering various options. One option is to print a selected range of pages. You can do this by typing the appropriate page numbers in the "From" and "To" boxes. When you're ready to print, click on the "OK" button.

Interrupting Calculations

If *Mathematica* is hung up in a calculation, or is doing something you don't want it to be doing, you can often free it up by typing "ALT+." (*i.e.*, hold down the ALT key and type a period). You can get the same effect by selecting Abort Evaluation from the Kernel menu. Either method should eventually cause *Mathematica* to stop what it's doing.

Ending the *Mathematica* Session

It is important to quit *Mathematica* when you are done working. First make sure that you have saved your work. Then select the Exit button from the File menu.

If you haven't saved all your work, *Mathematica* will give you one last chance to save it. When you have exited *Mathematica*, its menu and any open Notebooks will disappear.

Don't forget to eject your disk before you leave. On most machines, you eject the disk by pushing the button just below the slot for the disk.

Mathematica on a *Macintosh*

When you sit down at a *Macintosh* it may be turned off, or it may be running a screensaver program. If it's off, turn it on and wait until it is finished booting. If a screensaver is running, move the computer's mouse to refresh the display. If all goes well, along the top of the screen you should see the following (from left to right): an Apple icon; a series of menu buttons; the time; a "Help" icon indicated by a question mark; and in the upper right-hand corner, an "Application" icon. Click and hold on the Application icon. A menu will appear that will show, among other things, a list of applications running on the *Macintosh*. There will be a check mark on the button corresponding to the current application. Drag the mouse down to the "Finder" button, and then release. Your screen may change, depending on which application was active, and you should see one or more windows or panels on the screen. You are now in the Finder.

The Finder is the *Macintosh* program for finding, copying, moving, and organizing files. Each of the icons in the Finder represents either a file or a folder; a folder is a group of files bundled together. Clicking on a file or folder icon will *select* the icon, and the icon will be darkened. Double-clicking on a folder icon will cause the Finder to bring up a new window displaying a set of icons representing the files in that folder. Double-clicking on a file icon will open the file. The *Macintosh* tries to open a file using an appropriate *application*, or program. For example, *Mathematica* files will be opened using the *Mathematica* program. Double-clicking on such a file will open it in *Mathematica*. You can usually tell the type of a file by looking at its icon. For example, the icon for *Mathematica* files is a stellated icosahedron (a three-dimensional star).

If a window is partially obscured you can bring it to the front by clicking on any part of it. Clicking in the small box in the upper left-hand corner of a window will make the window disappear.

Moving files from one folder to another is done as follows: find the folder into which you want to move (or copy) a file. This folder can be either in iconic form, or in the form of an open window showing its contents. Then find the file you want to move and drag it to the target folder. An icon representing the file will appear in the target folder. (The file will be moved or copied, depending on who owns the file and folder.) To delete a file, drag its icon into the trash can at the bottom right-hand corner of the screen. Then select `Empty Trash...` from the `Special` menu at the top of the screen. A warning panel will appear on the screen asking you to confirm that you want to delete the files in the trash can.

Online Help for the *Macintosh*

If you have never used a *Macintosh* computer before, it is probably a good idea to look at the *Macintosh* "Tutorial" or the "Macintosh Guide", both of which are accessible from the Help menu. The tutorial will cover the basics of managing files and opening applications and will help you become comfortable using the *Macintosh*. You can also get a rudimentary kind of help by turning on the Show Balloons option in the Help menu.

Starting *Mathematica*

The *Mathematica* program uses a lot of memory, so before you start *Mathematica* you should quit any other applications that are running on the *Macintosh*. To do this, use the menu attached to the Application icon in the upper right-hand corner of the screen.

The icon for the *Mathematica* application is a stellated icosahedron with the name "Mathematica" printed below. It should be in a folder called "Mathematica". The location of this folder will depend on your particular machine, so you may have to hunt around in the Finder to locate it. Try looking in folders that have the word "Application" or "Math" in their titles. When you've found the icon, double-click on it to open it. After a brief delay, two new windows will appear on the screen, and the menu buttons along the top of the screen will change. The large window labeled "Untitled-1" is a *Notebook*. The smaller window to the right is a *Palette*, which can be used to insert mathematical symbols into the Notebook. You can now start typing in the Notebook.

Saving Your Notebook on a Floppy Disk

To save your Notebook onto a disk, you'll need to buy some disks for the computer you are using (usually 3.5 inch, 1.44 megabyte High Density (HD) disks). These are inexpensive and available in many book stores and computer stores. Insert a disk in the slot in the front of the *Macintosh*. If the disk is unformatted, or formatted for a different kind of computer, a panel will appear asking you whether you want to format it. If you click Initialize, a panel will appear asking whether you really want to erase the disk. If you click Erase, the *Macintosh* will prompt you for a name for the disk. Type in a name, then hit RETURN. It will take a few minutes to initialize the disk.

Once you have a properly formatted disk inserted in the drive, an icon representing the disk will appear along the right-hand edge of the screen. Make sure that your *Mathematica* window is active by clicking the mouse in it. A darkened titlebar indicates that the window is active. Now select Save As... from the File menu. A new panel will appear on the screen. Select the button Desktop on the right-hand side. In the center of the panel you should see the name of your disk. Click on the name, and then click Open. The panel will change, and there will be a box labeled "Save as:". Type the name you'd like to use for the Notebook ("Assignment 1", for example), and then click Save. A copy of the file will be saved on the disk, and the name of the main window will change from "Untitled-1" to the name you typed. To check that the file has been copied, double-click on the disk icon. A new window will appear containing a list of files on the disk. To continue working with *Mathematica*, click on your Notebook.

Once a Notebook has a title, you can save further changes by simply selecting Save from the File menu. *Mathematica* automatically keeps track of where you want to save it. If the disk on which you originally saved the file is not in the drive, the *Macintosh* will pop up a panel with an error message, or prompt you to insert the disk. In the latter case, you can get rid of the panel by typing "COMMAND+.". (The COMMAND keys are located on either side of the space bar, with both an apple and a cloverleaf-like symbol.) If you want to save the Notebook under a different name, you'll have to use the Save As... button again. You should save your work frequently.

If the computer you are using is networked, and you have an account on the network, you may be able to save your Notebook on a central file server. Consult your system administrator for instructions.

Opening a Previously Saved Notebook

To open a saved Notebook, click on the File button, and then click on the Open button. A new dialog panel will appear. This panel is essentially identical to the one that appeared as a consequence of selecting the Save As... button. You can navigate in this panel by clicking on drives and/or directories until you find the file you want. After selecting the file, click on the "Open" button.

To open a new Notebook, click the New button in the File menu. You can open and save as many Notebooks as you like, using a different name for each one, as long as you don't use up the available disk space.

If you have several Notebooks open, you may want to close one or more. To close a Notebook, make it active by clicking in it or by selecting its name from the menu labeled Window, and then select Close from the File menu. *Mathematica* will prompt you to save the Notebook if the Notebook has changed since the last save. Do not attempt to close a Notebook by selecting Exit from the File menu; that will terminate your *Mathematica* session entirely.

Printing Your Notebook

To print a Notebook, first make sure the Notebook you want to print is the active window. Then select Print... from the File menu. A new panel will appear offering various options. One option is to print a selected range of pages. You can do this by typing the appropriate page numbers in the "From" and "To" boxes. When you're ready to print, click on the Print button.

Interrupting Calculations

If *Mathematica* is hung up, or is doing something you don't want it to be doing, you can often free it up by typing "COMMAND+." (*i.e.*, hold down the COMMAND key and type a period). You can get the same effect by selecting `Abort Evaluation` from the `Kernel` menu. Either method should eventually cause *Mathematica* to stop what it's doing.

Ending the *Mathematica* Session

It is important to quit *Mathematica* when you are done working. First make sure that you have saved your work. Then select the `Exit` button from the `File` menu. If you haven't saved all your work, *Mathematica* will give you one last chance to save it. When you have exited *Mathematica*, its menu and any open Notebooks will disappear.

   Don't forget to eject your disk before you leave. You can eject the disk by selecting `Eject Disk` from the `Special` menu in the Finder, or by dragging the disk icon into the trash can.

Mathematica in the *X Window System*

We assume that you are using a computer running some version of the *UNIX* operating system and that you are using the *X Window System*. We also assume that you have an account on the computer and know how to log on. The *X Window System* is highly customizable, and you will have to learn the idiosyncrasies of your particular installation. In many installations, windows or panels will appear with a titlebar at the top, with buttons in the right- and left-hand corners. You can bring a partially obscured window to the front by clicking its titlebar. To do much more than this with your windows, you will have to consult site-specific documentation and/or your system manager.

   One of the windows on your screen should be a terminal window (probably titled "xterm" or "login"). You can type *UNIX* commands in this window. Here is a list of the basic *UNIX* commands for locating, moving, copying, and deleting files.

UNIX Command	Effect
ls	lists your files
cp file.a file.b	puts a copy of file.a into file.b
mv file.a file.b	renames file.a to file.b
rm file.a	deletes file.a

   In fact, you should be able to do most of the necessary file manipulation of your *Mathematica* files from within the *Mathematica* program (to be explained below), but on occasion you may need to maneuver your *Mathematica* files exactly as you would any other files in the *UNIX* environment.

Online Help for the *X Window System*

The usual way to obtain help in a *UNIX* environment is to consult the manual pages using the "man" command. For example, by typing the command "man ls" you can learn about the features of the "ls" command, including options that can be used with it. If you are a *UNIX* novice, you may find it instructive to try the "man" command on some of the basic system commands listed above.

Starting *Mathematica*

Mathematica is often started from an xterm window by typing "mathematica" at the *UNIX* prompt. Alternatively, *Mathematica* might be started on your system by clicking on a special menu button. You may need to ask your system administrator for the precise command or start-up procedure. Whatever the procedure, two new windows will appear on the screen. The large window labeled "Untitled-1" is a *Notebook*. The smaller window to the right is a *Palette*, which can be used to insert mathematical symbols into the Notebook. You can now start typing in the Notebook.

Saving Your Notebook

To save a Notebook, click on the `File` button in the menu row. A new menu will appear below the button. Click on the `Save As...` button. A panel will appear in the middle of the screen, containing a list of directories and *Mathematica* Notebook files. By double-clicking on a directory name, you cause the selected directory to become the current directory; the panel will change to show the subdirectories and files in the new current directory. If there are more entries in the current directory than will fit in the panel, you can use the mouse to manipulate the slider on the right of the panel to scroll up and down through the list of subdirectories.

   Once you have reached the directory in which you wish to save the Notebook (by double-clicking on successive subdirectories or typing a pathname into the Filename box) you should place the pointer at the end of the pathname in the selection box and click. Then type in the name you wish to give to your Notebook. The name should end with the suffix ".nb". Then click on the "OK" button. The panel will disappear, the Notebook will be saved, and its title will change from "Untitled-1" to the name you typed. If after working some more you want to save your Notebook again using the same name, all you have to do is click on the `Save` button instead of the `Save As...` button. If you want to save it under a new name, you must repeat the steps above. You should save your work frequently.

Opening a Previously Saved Notebook

To open a saved Notebook, click on the `File` button, and then click on the `Open` button. A panel that is essentially identical to the one that resulted from selecting the `Save As...` button will appear. You can navigate the directory tree by double-clicking on directories (or by typing a pathname into the "File name" region) until the file you want appears. Then double-click on the name of the appropriate file (or click on the name and select "OK").

   To open a new Notebook, click the `New` button in the `File` menu. You can open and save as many Notebooks as you like, using a different name for each one, as long as you don't use up the available disk space.

If you have several Notebooks open, you may want to close one or more. To close a Notebook, make it active by clicking in it, and then select `Close` from the `File` menu. *Mathematica* will prompt you to save the Notebook if the Notebook has changed since the last save. Do not attempt to close a Notebook by selecting `Exit` from the `File` menu; that will terminate your *Mathematica* session entirely.

Printing Your Notebook

To print a Notebook, select `Print` from the `File` menu. A panel will appear in the center of the screen. In the region labeled "Print Destination" there are two tiny buttons next to the words "File" and "Print to". If you click on the button next to "Print to", your Notebook will be directed to the printer. If you click on the button next to "File", the Notebook will be written to a PostScript file with suffix ".ps". The default filename will be shown to the right of the word "File". If you want to send the output to a PostScript file with a different filename, you can click on the `Browse...` button to bring up a File Browser panel, which you can use to select a new filename. Finally, when you have everything as you want it, click on the "OK" button at the bottom of the panel.

You will notice that there are several other options that you can select in the Print panel. In particular, you can select a range of pages. You do this by typing the appropriate page numbers in the "From" and "To" boxes.

The Mod1 Key

Somewhere on your keyboard is a key that we refer to as the Mod1 key. On many keyboards it is located next to the space bar, and may have a diamond or other graphic symbol on it. To locate the Mod1 key on your keyboard, click on the `X Environment Information` button in the `Help` menu. Then click on "Find Keys and Modifiers" in the panel that appears. You can learn the status of any key by typing it. Press a key that you think might be the Mod1 key and look at the description in the panel. When you find the right key, it will say "Mod1" next to the "Modifiers" label.

One important use for the Mod1 key is to interrupt calculations. If *Mathematica* is hung up in a calculation, you can often free it up by typing "Mod1+."; *i.e.*, hold down the Mod1 key and type a period. You can get the same effect by selecting `Abort Evaluation` from the `Kernel` menu. Either method should eventually cause *Mathematica* to stop what it's doing.

Ending the *Mathematica* Session

When you are done working, it is important to quit *Mathematica*. First make sure that you have saved your work. Then select the `Quit` button from the `File` menu. If you haven't saved all your work, *Mathematica* will give you one last chance to save it. When you exit *Mathematica*, any open Notebooks will disappear.

Chapter 3

Doing Mathematics with *Mathematica*

These instructions are designed to get you started doing mathematics with *Mathematica*. The software is fairly easy to use, and with the help of these instructions you should quickly master the basic skills necessary to work with it. Before you read this chapter you should already have read the appropriate section of Chapter 2. After you have read that section, you will know how to get *Mathematica* up and running. Then you should read this chapter, trying out the commands in a *Mathematica* Notebook as you go along. For further practice with these *Mathematica* commands, work the problems in Problem Set A.

Input and Output

You interact with *Mathematica* using "Notebooks". A Notebook appears on your computer screen as a window with various kinds of text and graphics in it. But *Mathematica* also treats a Notebook as a *document*, and in particular, *Mathematica* knows how to break the Notebook into pages to produce a finished, printed version of the Notebook. A Notebook is divided into *cells*, which are delineated by a bracket in the right-hand margin, and which can be of several different types. We'll discuss this cell structure in detail later, but in this section we discuss only two types of cells: *Input* cells and *Output* cells. An *Input* cell is a place where you type expressions for *Mathematica* to evaluate. An *Output* cell is a place where *Mathematica* gives its response.

The first thing to keep in mind when using *Mathematica* is that there is a difference between typing the ENTER key (just above the SHIFT key on the right-hand side) and typing SHIFT+ENTER (holding down the SHIFT key and pressing ENTER). Typing SHIFT+ENTER tells *Mathematica* to go ahead and evaluate whatever you've typed in. The ENTER key is a simple linefeed; it generates a new line without causing *Mathematica* to evaluate the input. It is useful if you want to group several commands or expressions together in a single *Input* cell. (On some systems, the ENTER key is labeled RETURN. On all systems, the ENTER key in the numeric keypad has the same effect as SHIFT+ENTER.)

As an example, try typing **1 + 1**. It should appear at the top of the Notebook. To evaluate this expression, press SHIFT+ENTER. Notice that while *Mathematica* is working, it displays the word "Running..." at the top of the Notebook. It will take some time for *Mathematica* to answer because *Mathematica* completes its loading process upon the first calculation. From this point, responses to simple inputs should be nearly instantaneous.

Your Notebook should now look something like this:

```
In[1]:=  1 + 1

Out[1]=  2
```

The symbol *In[1]:=* is an input label in *Mathematica*, and it appears (with a different line number) at the beginning of every *Input* cell after the cell has been evaluated. The output of the command is in an *Output* cell and has the label *Out[1]=*. Material in the *Output* cells is specially formatted in mathematical notation.

Arithmetic

As we have just seen, *Mathematica* can do arithmetic like a calculator. You can add with "+", subtract with "−", multiply with "*", divide with "/", and exponentiate with "^". For example,

```
In[2]:=  2*5 + 3^2 - 4/2
Out[2]=  17
```

```
In[3]:=  (5 + 3)(5 - 3)
Out[3]=  16
```

Notice in the last example that the "*" for multiplication was left out. You can always use a space instead of "*", even between two numbers: **2 5** would evaluate to 10 just like **2*5**. In some cases, such as between parentheses, or in the case of a number times a variable, even the space can be left out.

Mathematica differs from a calculator in that it treats fractions symbolically rather than converting to decimal approximations. For example, if you type in **4/7**, *Mathematica* will simply respond "$\frac{4}{7}$". To force *Mathematica* to give you a numerical answer, type **N[4/7]**.

```
In[4]:=  N[4/7]
Out[4]=  0.571429
```

To get 20 digits rather than 6 in the answer, you would type **N[4/7, 20]**. Another way to force *Mathematica* to use decimal approximations rather than exact numbers is to use a decimal point. For example, in *Mathematica*, "**2.0**" is different from "**2**". It is important to remember this distinction because approximate, or floating point, arithmetic is usually faster than exact arithmetic. Floating point output is often less complicated as well. To see an example of the difference in speed and appearance between floating point and exact arithmetic, evaluate the following two commands in your Notebook.

```
Solve[x^8 - x^2 - 2 == 0, x]

Solve[x^8 - x^2 - 2.0 == 0, x]
```

Errors in Input

If you make an error in an input line, *Mathematica* will print an error message. For example,

```
In[5]:= (5 + 3 (5 - 3)
```

```
        Syntax::sntxi: Incomplete expression;
                       more input is needed.
```

Although *Mathematica* cannot always tell you exactly what is wrong, you should pay attention to error and warning messages. Missing brackets and parentheses are the most common errors on input lines.

You can edit an input line by using the mouse to position the cursor at the desired insertion point, clicking the button on the mouse, then typing (or using the "delete" or "backspace" keys), and finally pressing SHIFT+ENTER to reenter the line. You can also use the arrow keys to move around in the Notebook.

Online Help

To access the online documentation for a *Mathematica* command, type **?command** in an *Input* cell. For example, the input **?FindRoot** will yield information on the **FindRoot** command. If you can't remember the name of a command, you might try looking for it using the browser feature of the Notebook help facility. Alternatively, you could guess at all or part of the name of the command, and use asterisks to check for all commands containing the string you typed. For example, to get a list of all *Mathematica* commands starting with "F", you would type **?F***. Or if you want to factor a number, but don't know the name of the appropriate command, you could try typing **?*Factor*** to see a list of *Mathematica* commands containing the word "Factor".

Mathematica's online help also includes the Help Browser, which contains more extensive information about and examples of commands, options, and packages. To open the Help Browser, choose Help... from the Help menu. To go directly to the information about a particular command, type its name in the box at the top of the window and then press the "Go To" button. Alternatively, you can click the mouse button on the list of topics in the center panels of the window to explore related commands.

Many *Mathematica* commands can be modified by using options. An option is specified with the syntax **OptionName** $\rightarrow$ **OptionValue** (the arrow is typed as a minus sign "**-**" followed by a greater than sign "**>**"). We will see an example in the *Graphics* section near the end of this chapter. You can see a list of all possible options to a command, and their default values, by typing double question marks before the command name, as in **??FindRoot**. You can then get more information about an option by typing a question mark before the option name. For example, **?Direction** produces information about **Direction**, one of the options for the **Limit** command.

In most cases, options are fairly technical, and you will not have to modify them. But in the plotting commands many of the options are quite useful.

Algebra

You can add, subtract, multiply, and divide variables as well as numbers. Consider the following series of examples:

In[6]:= **(x + y)(x + y)^2**

Out[6]= $(x + y)^3$

In[7]:= **Expand[%]**

Out[7]= $x^3 + 3x^2y + 3xy^2 + y^3$

In[8]:= **Factor[%]**

Out[8]= $(x + y)^3$

(The "**%**" symbol refers to the output of the previous command; see the section *Referring to Previous Output* later in this chapter.) *Mathematica* often makes minor simplifications to the expressions you type in, but does not make any major changes unless you tell it to. The **Expand** command forced *Mathematica* to multiply the expression out, and the **Factor** command told *Mathematica* to put it back in factored form.

Mathematica also has a **Simplify** command, which tries to express a formula as simply as possible. Sometimes this involves factoring, and sometimes multiplying things out. For example,

In[9]:= **Simplify[(x^2 - y^2)/(x - y)]**

Out[9]= $x + y$

In[10]:= **Simplify[(x + y)(x^2 - x y + y^2)]**

Out[10]= $x^3 + y^3$

Notice the space between **x** and **y** on the last input line. This space is necessary—*Mathematica* would have interpreted **xy** as a new variable rather than the product of **x** and **y**. Even better, we could put a "*****" between **x** and **y**.

Equations and Assignments

In *Mathematica*, a single equal sign is used to assign values to a variable. For instance,

```
In[11]:= x = 5

Out[11]= 5
```

will give the variable **x** the value 5 from now on. Whenever *Mathematica* sees an **x**, it will substitute the value 5.

```
In[12]:= x^2 + 3 x*y + y

Out[12]= 25 + 16y
```

To clear the variable **x**, type either **x = .** or **Clear[x]**.
Assignments can be quite general.

```
In[13]:= Clear[x]

In[14]:= z = x^2 + 3 x*y + y

Out[14]= x^2 + y + 3xy

In[15]:= z + 7 y

Out[15]= x^2 + 8y + 3xy

In[16]:= y = 5

Out[16]= 5

In[17]:= z

Out[17]= 5 + 15x + x^2
```

A variable or function name can be any string of letters and digits, as long as it begins with a letter. You should choose names that are both distinctive and easy to remember. For example, you might use **solndampedpend** as the name of the solution of a differential equation that models a damped pendulum.

A common source of puzzling errors in a *Mathematica* session is to forget that you have defined variables. (*Mathematica* never forgets.) You can always check on the current value of a variable by typing a question mark followed by the variable name. For example, to find the current value of the variable **z**, type **?z**.

Equations in *Mathematica* are indicated by a double equal sign. For example, to find the root of the equation $e^x = x + 2$ that is close to $x = 1$, we type

In[18]:= **FindRoot[Exp[x] == x + 2, {x, 1}]**

Out[18]= $\{x \to 1.14619\}$

Mathematica can also solve sets of equations.

In[19]:= **Solve[{r + s == 3, r - s == 0}, {r, s}]**

Out[19]= $\left\{\left\{r \to \dfrac{3}{2}, s \to \dfrac{3}{2}\right\}\right\}$

Built-in Functions

Mathematica has the usual functions from calculus and differential equations already "built-in." These include **Sqrt**, **Cos**, **Sin**, **Tan**, **Log**, and **Exp**. *Mathematica* has many other built-in functions, including less familiar mathematical functions like **Erf**, **Gamma**, and **BesselJ**. *Mathematica* also has a few built-in constants, such as **Pi** (the number π), **E** (the base e of the natural logarithm), **I** (the complex number $i = \sqrt{-1}$), and **Infinity** (∞). Some examples:

In[20]:= **Log[E^2]**

Out[20]= 2

In[21]:= **Sin[Pi/3]**

Out[21]= $\dfrac{\sqrt{3}}{2}$

(To get numerical answers one would use the **N** function.) Notice that **Log** is the natural logarithm, called *ln* in many texts. Note also the use of square brackets instead of parentheses. *Mathematica* would interpret the input **Log(E^2)** as meaning a variable called "Log" times the number e^2. And finally, notice that all of *Mathematica*'s commands, built-in functions and constants begin with a capital letter. This is to avoid confusion with any variables or functions you might define; you should generally stick to lowercase letters for things you define.

User-defined Functions and Expressions

It is possible to define new functions in *Mathematica*. In the following example, a polynomial function is defined and evaluated at one point.

In[22]:= **f[x_] := x^3 + 7x - 5**

In[23]:= **f[2]**

Out[23]= 17

Notice the underscore after the initial **x**, which means that **x** is a "dummy variable" in the definition, and the colon before the equals sign, which tells *Mathematica* not to evaluate the right-hand side until you actually use the function.

Another kind of object in *Mathematica*, similar to a function, is an *expression*. For example, **x^3 + 7x - 5** is an expression. We can assign an expression to a variable, as follows:

```
In[24]:= g := x^3 + 7x - 5
```

Although this *expression* looks very much like the *function* we defined above, it is in fact quite different. The function is a rule for changing x into some other quantity, while the expression is merely a quantity involving x. In particular, typing **g[2]** will not give the value of the expression at $x = 2$. (Try it to see what happens!) We can, however, evaluate an expression at a particular point by using the replacement operator "**/.**" and a transformation rule.

```
In[25]:= g /. x -> 2
```
```
Out[25]= 17
```

This input means: replace x by 2 in the expression g.

Referring to Previous Output

Mathematica labels every input and every output; this makes it easy to use previous results in future calculations. The "**%**" symbol means "the last output", "**%%**" means "the next-to-last output", and **%n** refers to the output labeled *Out[n]*. Here is an illustration:

```
In[26]:= 2 + 2
```
```
Out[26]= 4
```

```
In[27]:= %^2
```
```
Out[27]= 16
```

```
In[28]:= %% * 5
```
```
Out[28]= 20
```

```
In[29]:= %26/4
```
```
Out[29]= 1
```

A safer way to refer to past output is to assign the output to a variable. For example, we could type:

In[30]:= **total = 2 + 2**

Out[30]= 4

In[31]:= **total^2**

Out[31]= 16

The reason this is safer is that in the Notebook, the output that is in the preceding *Output* cell is not necessarily the most recent output generated by *Mathematica*. Even more problematic is the situation in which there are multiple Notebooks open in a single *Mathematica* session. If you switch between Notebooks, the preceding output of *Mathematica* may not even be in the current Notebook. It is a good idea to assign names to any output you expect to use later in a session.

Lists and Tables

Mathematica arranges its data in the form of lists. A list can contain any collection of *Mathematica* objects, including numbers, variables, functions, and equations. Lists are entered in the form **{a, b, c}**, where **a**, **b**, and **c** are the elements of the list. In many cases, lists can be treated exactly like single objects. For example, we could type

In[32]:= **v = {1, 2, 3, 4, 5}**

Out[32]= $\{1, 2, 3, 4, 5\}$

In[33]:= **v^2**

Out[33]= $\{1, 4, 9, 16, 25\}$

In effect, we instructed *Mathematica* to square the list **v**, and it responded by squaring each element of the list. You can also add, multiply, subtract, and divide lists, plot lists of functions, and solve lists of equations. You can even apply functions to lists, as in **f[v]**.

To extract the nth element of a list **v**, type **v[[n]]** or **Part[v, n]**. The **First** and **Last** commands can be used to extract the first and last elements of a list, respectively, as in **First[v]**. A sublist of elements can be extracted from a list by specifying a list of positions. For example, **v^2[[{2, 4, 5}]]** or **Part[v^2,{2, 4, 5}]** produces a list consisting of the second, fourth, and fifth elements of **v^2**.

As we shall see later, the **First** command is especially useful for extracting solutions to differential equations generated by the *Mathematica* commands **DSolve** and **NDSolve**.

The **Join** command is used for combining lists. For example,

```
In[34]:= list1 = {1, 5, 10}
         list2 = {9, 5, 0}
         Join[list1, list2]
```

Out[34]= $\{1, 5, 10\}$

Out[35]= $\{9, 5, 0\}$

Out[36]= $\{1, 5, 10, 9, 5, 0\}$

Note that we have entered three lines in a single *Input* cell by using the RETURN key.

An easy way to generate lists is to use *Mathematica's* **Table** command. For example, to generate a table of values of the function **f** (defined above) on the even integers from 0 to 10 you would type

```
In[37]:= Table[f[x], {x, 0, 10, 2}]
```

Out[37]= $\{-5, 17, 87, 253, 563, 1065\}$

In this example, the "**0**" represents the starting value of **x**, "**10**" is the ending value, and "**2**" is the increment (so we get only even integers in this case). The increment is an optional argument, the default increment being 1. The output of the **Table** command can be put in tabular form by appending the command **//TableForm**. In the following example, we make a table of values of **x** together with values of **f[x]**, for integer values of **x** from -1 to 3.

```
In[38]:= Table[{x, f[x]}, {x, -1, 3}]//TableForm
```

Out[38]//TableForm=
$$
\begin{array}{rr}
-1 & -13 \\
0 & -5 \\
1 & 3 \\
2 & 17 \\
3 & 43
\end{array}
$$

In certain situations you may want to construct lists that depend on two or more parameters. You might do this, for example, if you're solving a second order differential equation in y, and want to study the solutions for various values of the initial conditions $y(x_0) = y_0$, $y'(x_0) = y'_0$. You can do this by using multiple parameters in the **Table** command. Here is an example in which we construct the set of all fractions with numerator and denominator between 1 and 3.

```
In[39]:= Table[a/b, {a, 1, 3}, {b, 1, 3}]
```

Out[39]= $\left\{ \left\{ 1, \frac{1}{2}, \frac{1}{3} \right\}, \left\{ 2, 1, \frac{2}{3} \right\}, \left\{ 3, \frac{3}{2}, 1 \right\} \right\}$

Note that there are two levels of braces, one for each parameter in the **Table** command. To remove the extra braces we can type

In[40]:= **Flatten[%]**

Out[40]= $\left\{1, \dfrac{1}{2}, \dfrac{1}{3}, 2, 1, \dfrac{2}{3}, 3, \dfrac{3}{2}, 1\right\}$

Finally, to eliminate all repetitions from this list we would type

In[41]:= **Union[%]**

Out[41]= $\left\{\dfrac{1}{3}, \dfrac{1}{2}, \dfrac{2}{3}, 1, \dfrac{3}{2}, 2, 3\right\}$

Graphics

Here is the basic command for plotting an expression involving one variable. We will encounter several other plotting commands as we go along.

In[42]:= **Plot[x^3 - x, {x, -1, 1}]**

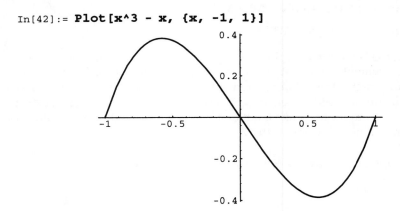

Out[42]= —Graphics—

Plots can be misleading if you do not pay attention to the axes, which are not always what you'd expect. For example, the input **Plot[x^3 - x + 1, {x, -1, 1}]** will produce a plot that looks identical to the previous one, except that the horizontal axis is no longer the x-axis, but rather the horizontal line through $y = 1$.

To produce a plot with a specified range, use the **PlotRange** option, as in:

Plot[x^3 - x + 1, {x, -1, 1}, PlotRange -> {-2, 2}]

This produces a graph with horizontal range $[-1, 1]$ and vertical range $[-2, 2]$.

By default, *Mathematica* uses different scales on its horizontal and vertical axes. It chooses the scales so that the ratio of height to width in the final graph is the inverse of the global ratio. You can force *Mathematica* to use the same scale on both axes by setting **AspectRatio -> Automatic**.

Aborting Calculations

If *Mathematica* gets hung up in a calculation, or seems to be taking too long, you can usually stop it by pressing "ALT+."; *i.e.* hold down the ALT key and then press the period key. (In the *X Window System*, press "Mod1+."; on the *Macintosh*, press "COMMAND+.". See Chapter 2.) It may take a while for *Mathematica* to abort the calculation, as it tries to clean up the memory it has used during the calculation. If *Mathematica* is hung up, you should try to save your work using the menu button, and then quit *Mathematica* and restart it. It is a good idea when using any computer program to save your work often, as an unexpected crash or power outage can wipe out all your unsaved work.

Packages

The *Mathematica* system actually consists of several distinct parts. The part that reads input, displays output, and manipulates Notebooks is called the *interface*, or *front end*. The computational engine, the part that actually evaluates the input, is called the *kernel*. The kernel contains the core parts of *Mathematica* that must always be available. A third part consists of *Mathematica packages*, which contain additional commands. To use commands in packages, you must explicitly tell *Mathematica* to load the commands. For example, the **PlotVectorField** command is part of the **Graphics`PlotField`** package, and must be loaded before it can be used. To load this package, type

```
In[43]:= <<Graphics`PlotField`
```

(Note that the word **PlotField** is surrounded by *backwards* quotes.) If you use a package command without loading it, *Mathematica* will not recognize the command and will simply echo the input line in the output area. If you do make this mistake, you can recover by typing **Remove[command]** and then loading the package properly. For example, if we had tried to use the **PlotVectorField** command before loading it, we would have to type **Remove[PlotVectorField]** before we could properly load the command and use it.

The two packages you will need most often in doing the problem sets are called **Graphics`PlotField`** and **Calculus`LaplaceTransform`**. The former contains a collection of specialized routines for graphing vector fields. The latter contains routines for working with Laplace Transforms.

Problem Set A

Practice with *Mathematica*

In this problem set, you will use *Mathematica* to do some basic calculations, and then to plot, differentiate and integrate various functions. This problem set is the minimum you should do in order to reach a level of proficiency that will enable you to use *Mathematica* throughout the course. In this problem set you should concentrate on the mathematics, but if you wish to try out some of the editing features of the Notebook interface, you can read about them in the next chapter.

1. Evaluate:

    (a) $3 + 9$

    (b) 2^{123}

    (c) π^2 and e to 35 digits

    (d) the fractions $\frac{22}{7}$, $\frac{311}{99}$, and $\frac{355}{113}$, and determine which is the best approximation to π.

2. Evaluate to ten digits:

    (a) $\dfrac{\sin(0.1)}{0.1}$

    (b) $\dfrac{\sin(0.01)}{0.01}$

    (c) $\dfrac{\sin(0.0001)}{0.0001}$.

3. Factor the polynomial $x^3 - y^3$.

4. Use the **Plot** command to graph the following:

    (a) $y = 3x + 2$ for $-5 \le x \le 5$

    (b) $y = x^2 + x - 1$ for $-5 \le x \le 5$

    (c) $y = \sin x$ for $0 \le x \le 4\pi$

    (d) $y = \tan x$ for $-\pi/2 \le x \le \pi/2$

(e) $y = e^{-x^2}$ for $-2 \le x \le 2$.

5. Plot the functions x^4 and 2^x on the same graph, and determine how many times their graphs intersect. (*Hint*: You will probably have to make several plots, using intervals of various sizes, in order to find all the intersection points. You may have to limit the y-range in your plots.) Now find the values of the points of intersection using the **FindRoot** command.

6. The **ContourPlot** command takes a function of two variables (say $f(x, y)$) and plots several *level curves* of the function (that is, curves of the form $f(x, y) = c$, where c is a constant). For example, **ContourPlot[x^2 + y^2, {x, -5, 5}, {y, -5, 5}]** plots the level curves of the function $f(x, y) = x^2 + y^2$, in the square $-5 \le x \le 5, -5 \le y \le 5$.

   (a) Use **ContourPlot** to plot level curves of the function $f(x, y) = 3y + y^3 - x^3$, in the region where x and y are between -1 and 1 (to get an idea of what the curves look like near the origin) and in a larger region (to get the big picture).

   (b) A particular level set of a function can be plotted using **ContourPlot** with the options **Contours -> {c}**, **ContourShading -> False**, where "c" is the value of the constant corresponding to the particular level curve to be plotted. Plot the curve $3y + y^3 - x^3 = 5$.

   (c) Plot the level set of the function $f(x, y) = y \ln x + x \ln y$ that contains the point $(1, 1)$. (You will have to compute the appropriate value of "c" by evaluating the function at the point $(1, 1)$. Note that the logarithm is only defined for positive values of x and y.)

7. The derivative of a function can be found by typing **D[f[x], x]**. Find the derivatives of the following functions. In each case, if it appears that the answer *Mathematica* gives can be simplified, investigate the possibility by using **Simplify**.

   (a) $f(x) = 7x^3 + 3x^2 - 2x + 1$

   (b) $f(x) = \dfrac{x + 1}{x^2 + 1}$

   (c) $f(x) = \cos(x^2 + 1)$

   (d) $f(x) = \arcsin(2x + 3)$

   (e) $f(x) = \sqrt{1 + x^4}$

   (f) $f(x) = x^r$

   (g) $f(x) = \arctan(x)$.

8. Consider the polynomial $x^6 - 21x^5 + 175x^4 - 735x^3 + 1624x^2 - 1764x + 720$.

   (a) Factor it.

(b) Find the roots using the *Mathematica* command **Solve**.

(c) Plot the polynomial for $0.5 \le x \le 6.5$.

(d) Plot the polynomial and its derivative on the same interval. Explain your plot using calculus.

9. Use the *Mathematica* command **Limit** to evaluate the following limits (in (d), you will have to use the **Direction** option):

(a) $\lim\limits_{x \to 0} \dfrac{\sin x}{x}$

(b) $\lim\limits_{x \to \pi} \dfrac{1 + \cos x}{x - \pi}$

(c) $\lim\limits_{x \to \infty} x\,e^{-x}$

(d) $\lim\limits_{x \to 1^-} \dfrac{1}{x - 1}$

(e) $\lim\limits_{x \to 0} \sin\left(\dfrac{1}{x}\right)$.

10. The integration command in *Mathematica* is **Integrate**. Type **?Integrate** to see the general syntax for both indefinite and definite integration. As an example, **Integrate[Cos[x], x]** gives the output Sin[x]. Notice that the constant of integration is left out; *Mathematica* gives you one possible antiderivative of the function you specified. As an example of the definite integral, the command **Integrate[Cos[x], {x, 0, Pi/2}]** gives the answer 1.

See if *Mathematica* can do the following integrals. For the indefinite integrals, check the results by differentiating:

(a) $\int_0^{\pi/2} \sin x \, dx$

(b) $\int x \cos(x^2) \, dx$

(c) $\int \sin(3x)\sqrt{1 - \cos(3x)} \, dx$

(d) $\int \ln x \, dx$

(e) $\int x^2 \sqrt{x + 4} \, dx$

(f) $\int \sqrt{x^4 + 1} \, dx$

(g) $\int e^{\cos x} \, dx$

(h) $\int_{-\infty}^{\infty} e^{-x^2} \, dx$.

11. *Mathematica* can also integrate functions numerically, to get an approximate answer. This is useful in cases where no elementary formula exists for an antiderivative, or where *Mathematica* cannot perform the symbolic integration. The command that does this is **NIntegrate**, which has the same syntax as **Integrate**. Find numerical values for the following integrals:

(a) $\int_0^\pi e^{\cos x}\,dx$

(b) $\int_0^1 \sqrt{x^4 + 1}\,dx$

(c) $\int_{-\infty}^\infty e^{-x^2}\,dx$.

For part (c), figure out how much the numerical answer differs from the exact answer found in the previous problem.

Chapter 4

Using *Mathematica* Notebooks

In this chapter, we describe some aspects of the Notebook feature of *Mathematica* Version 3.0 (many aspects are different in earlier versions). In particular, we describe some of the formatting tools that will help you create an attractive and polished document when you prepare solutions to the problem sets. Most of the Notebook features are the same on all platforms; we point out the relevant differences.

Since we refer frequently to menu items, we establish the following convention. When referring to an item in a submenu of another menu, we type the name of the top-level menu, then a colon, then the name of the item or submenu, and so on. For example, `Format:Face:Italic` means the `Italic` item of the `Face` submenu of the `Format` menu. To get to this item, first use the mouse to move the cursor to `Format` in the main menu. Click and hold down the left mouse button, and in the submenu that appears, move the cursor down to `Face`. Then move rightward onto the new submenu, down to `Italic`, and release the mouse button.

Keyboard Equivalents

You may have noticed that some menu items have not only a name but also some extra symbols on the right. These symbols are the *keyboard equivalent* for the given menu item. You may find these keyboard equivalents to be efficient alternatives to using the mouse. For example, the `Format:Face:Italic` item is followed in *Windows 95* by "Ctrl+I". This means you can hold down the *control key*, labeled CTRL on most keyboards, and press the "i" key to change text into or out of italics, just as if you had selected the corresponding menu item with the mouse. (Do not hold down the SHIFT key while typing a keyboard equivalent unless the word "Shift" appears next to the menu item; in *Windows 95* holding down SHIFT while typing CTRL+I will execute `Cell:Convert to:InputForm` instead.)

Many keyboard equivalents differ from one type of computer to another. Usually the difference is in which key to hold down while typing another key. See Chapter 2 for tips on learning the necessary keys for the *Macintosh* or the *X Window System*.

Manipulating Notebooks

Most Notebooks will be longer than the window on the screen. To scroll through the Notebook, drag the slider up and down in the scroll bar at the side of the window or use the "Page Up" and "Page Down" keys on your keyboard.

You can open more than one Notebook at a time by using the File menu. The New item will open a new untitled Notebook, while Open can be used to open a previously saved Notebook. Select Close to close a Notebook without quitting *Mathematica*. If you try to close an unsaved Notebook, *Mathematica* will ask whether you want to save it first.

The active Notebook is the one with the darkened titlebar, and anything you type will go into that Notebook. You can select a Notebook by clicking anywhere in its window (in the *X Window System* you may need to click on its titlebar to bring it to the front). To move a Notebook, drag its titlebar.

One reason to open more than one notebook is to copy and paste text from one Notebook to another. Note, however, that any definitions of variables, *etc.*, that you make in one Notebook carry over to every other open Notebook, so if you evaluate Input cells in two different Notebooks, you may get unexpected results.

Save your Notebooks often. This is a general principle when working with computers, because a sudden crash of the system could cause your unsaved work to be lost. The simplest way to save is to type the keyboard equivalent for File: Save (in *Windows 95*, CTRL+S) every 10 minutes or so.

Cells

In Chapter 3, we said that *Mathematica* Notebooks are organized into *cells*. There are several different types of cells, including *Input, Output, Graphics, Text, Title*, and *Section* cells. Input cells contain commands to be evaluated by *Mathematica*, and their output appears in Output and Graphics cells. You can use Title, Section, and Text cells to organize and insert commentary into your Notebooks. For example, to make and interpret a graph, you would first type a plotting command in an Input cell, then *Mathematica* would create the graph in a Graphics cell, and finally you would describe it in a Text cell.

Each cell type has a particular font and a particular type of formatting. For example, a Title cell has a large boldface font. A cell is delineated by a bracket along the right side of the Notebook. The easiest way to change the type of a cell is to select the cell by clicking on its bracket (the bracket will be highlighted), and then type a keyboard equivalent to change the type of the cell. For example, to change a cell to a Text cell, click on the cell bracket, and then type ALT+7. Nine common cell types have keyboard equivalents, which are as follows in *Windows 95*.

ALT+1	Title	ALT+4	Section	ALT+7	Text
ALT+2	Subtitle	ALT+5	Subsection	ALT+8	Small text
ALT+3	Subsubtitle	ALT+6	Subsubsection	ALT+9	Input

Another way to change the cell type is to use an item from the `Format:Style` menu. To see the type of a cell, select this menu and the type of the current cell will be marked in the resulting submenu. You can select another type from the submenu, or remind yourself of its keyboard equivalent.

Cursors

Mathematica uses a variety of cursors to indicate the different actions that are available at various places in the Notebook. Move the cursor up and down through a variety of cells and watch how it changes. We will describe the most important cursors. You can experiment with the others.

A horizontal I-bar cursor appears at the boundary between two cells or at the end of the Notebook. Clicking the mouse button when the cursor is a horizontal I-bar causes a horizontal line to be inserted into the Notebook. This horizontal line is a *cell insertion point*. To insert a new cell you select a cell insertion point, use an item from the `Format:Style` menu or a keyboard equivalent to select a cell type, and then start typing. For example, to insert a new Text cell at the end of a Notebook, you click at the end of the Notebook, type ALT+7 or select `Format:Style:Text`, and then start typing.

A vertical I-bar cursor appears when pointing at text that can be edited, *e.g.*, text in an Input or Text cell. Clicking will create a thin vertical line that indicates a *text insertion point*, where you can add text by typing, or delete text using the backspace or delete key. You can also use the arrow keys on your keyboard to move the insertion point around. Note that the insertion point, *i.e.*, the place where text appears, is generally different from the mouse cursor position.

In addition, there are several kinds of cursors that appear in Graphics cells, which are described below.

Cell Hierarchy

A Notebook is organized by grouping together blocks of cells. An example of this is the automatic grouping of Input cells with Output cells. A cell grouping is indicated by an extra bracket along the right edge of the Notebook that surrounds one or more cell brackets. When you use sectioning cells, the remaining cells are automatically grouped according to a *cell hierarchy*. For example, if you create a Title cell at the top of your Notebook, there will be an infinitely extendible cell bracket at the far right of the Notebook that will enclose every other cell you create in the Notebook. If you then create a Section cell, there will be a new extendible bracket created just inside the first one, which will surround all successive cells until you create another Section cell. The same is true for Subsection and Subsubsection cells.

Here is an example of how this can be useful. When you start the solution to a problem, say Problem 1, create a Section cell and type "Problem 1". Now create your solution to Problem 1 using Text cells, Input cells, Output cells, *etc.* All of these cells will be grouped together with a bracket along the right side of the

Notebook. If Problem 1 has parts (a), (b), *etc.*, then use a Subsection cell to denote the beginning of each part. When you are finished with Problem 1, create a new Section cell and type "Problem 2". A new grouping bracket for Problem 2 will be created, and subsequent material in the Notebook will be grouped in this bracket until a new sectioning cell is introduced.

These grouping brackets can be used to manipulate entire blocks of cells in the Notebook. For example, if you have created a section that contains the solution to Problem 1, you can click on its grouping bracket. The bracket will be highlighted and you can move the slider at the side of the Notebook up and down to verify that all the cells in the solution to Problem 1 are grouped together. If you press SHIFT+ENTER, then *Mathematica* will evaluate all the Input cells in the section, in order. If you double-click on the grouping bracket, the entire section will close up, and only its title will be visible. A small triangle at the bottom of the bracket indicates that its cell group is closed. To open the group, just double-click on its bracket once more.

If you experiment with this feature, you'll soon see how useful it is for organizing and expediting your work. Note that when you print the Notebook, it will appear much like it does on the screen. In particular, closed cell groups will print in closed form.

Manipulating Cells

Occasionally, you will want to combine or divide cells. To divide a cell, first put the insertion point where you want the division to occur, and then select Cell:Divide Cell. To merge several contiguous cells, first select the cells by dragging the mouse over a range of cell brackets, and then select Cell:Merge Cell. The selected cells will merge, and the cell type of the new cell will be the type of the topmost cell in the selection. Formatted cells, including Graphics and Output cells, generally should not be merged or divided.

The command Input:Copy Input from Above is useful for manipulating input. This command places a copy of the contents of the most recent Input cell at the current insertion point. This is especially useful if you want to execute several variants of a single input expression.

Editing

The basic editing commands are in the Edit menu. The most useful of these are Cut, Copy, and Paste. (Note the keyboard equivalents listed next to these items.) You select text to cut or copy by depressing the mouse button at one end of the selection, and holding it down as you move to the other end. The selected text will be highlighted. While the text is selected, choose either Cut or Copy. The former will delete the selection and save a temporary copy in a buffer. The latter will save a temporary copy without deleting anything. To paste text that has been cut or copied, move the cursor to the desired insertion point, click to set the insertion point, and choose Paste. Note that you can select an entire cell by clicking on its cell bracket, and the Cut, Copy, and Paste commands work for cells.

Graphics

You can use the Notebook interface to adjust and customize graphics. Click any-where inside a Graphics cell to select it. You can then move the graph by dragging it with the mouse. You can resize the graph by dragging one of the small squares along the edges that appear after you have selected the graph. If you wish to return your Graphics cell to its default size and alignment, select the cell (by clicking on its bracket), and choose `Cell:Make Standard Size`.

Using the mouse, you can determine the coordinates of points in a graph. To do this, select the graph, and hold down the CTRL key (the COMMAND key on the *Macintosh*; the "Mod1" key in the *X Window System*) while moving the mouse around. The cursor will change to crosshairs, and the coordinates of the point in the middle of the crosshairs will appear at the bottom of the Notebook window.

Online Help

Mathematica has extensive online help that you can access in several ways. First, you can get information about a *Mathematica* command by entering **?Command** or **??Command** in an Input cell. For example, try typing **?Plot** to get a description, including the syntax, of the **Plot** command. Typing **??Plot** yields more infor-mation, including the possible options to the **Plot** command and their default values.

Second, you can get more extensive information about *Mathematica* commands and the Notebook interface by using the Help Browser. To open the Help Browser, select `Help:Help....` There are six databases, listed near the top of the Help Browser, that you can select. (Some of the documentation may not be installed in your configuration.) The default database is "Built-in Functions"; another one that is very useful is "Master Index", which includes "Built-in Functions", "Add-ons", and "The Mathematica Book".

You can browse the selected database through the menus below the list of databases, or by typing the command name or topic you want help on in the "Go To" box. In the latter case, once you have typed enough that the topic you want appears in one of the menu boxes, press ENTER or click on "Go To". The help text will then appear in the bottom half of the Help Browser. Small pieces of the text may be underlined and colored blue; these indicate links that you can click on to jump to other sections of help text.

The "Other Information" database describes the Notebook interface, in partic-ular, all menu items. The "Add-ons" database includes information on packages of *Mathematica* commands that can be loaded in addition to the built-in func-tions. Finally, to learn more about using the Help Browser, select the "Getting Started/Demos" database and click on "Using the Help Browser".

Mathematical Typing

When inserting commentary in Notebooks it is often useful to be able to type mathematical formulas, symbols, subscripts, *etc.*, and to use different fonts in some cases. You can do these things through keyboard sequences, menu items, and *palettes*. The standard palette is the thin window that generally appears automatically when you start *Mathematica*; if no such window appears for you, then select `File:Palettes:BasicInput` to open it. Other palettes are also available in the `File:Palettes` menu, but we will concentrate on the BasicInput Palette, which offers common mathematical symbols, Greek letters, and templates for constructing formulas. Click on a button to insert the corresponding symbol, letter, or template into the active Notebook.

To type a formula in a Text cell, start by typing CTRL+9 (this is the keyboard equivalent for `Edit:Expression Input:Start Inline Cell`). Then type the formula using the keyboard and the palette. Many forms of mathematical notation, such as subscripts, superscripts, fractions, and square roots can be entered either from the keyboard or through a template from the palette. For example, to type x^2 you can type **x**, then CTRL+6, then **2**, then CTRL+SPACE (hold down the CTRL key and press the space bar). Alternatively, click on the superscript template in the upper left corner of the palette. A pair of boxes appear, waiting for you to type the appropriate symbols into them. Type **x**, then press TAB (or click in the superscript box with the mouse), then type **2**, and finally type CTRL+SPACE. Then continue typing your formula. When you are done with the formula, type CTRL+0 (for `Edit:Expression Input:End Inline Cell`), and continue typing text.

You can learn the keyboard equivalents for many common mathematical notations from the `Edit:Expression Input` menu. For more information on mathematical typing, select the "Other Information" database in the Help Browser, and then click on "2D Expression Input".

To change the font in a Text cell, use the items in the middle of the `Format` menu. For instance, to change a word, phrase, or entire cell into italics, use the mouse to select the text that you want to change. Then select `Format:Face:Italic` or type its keyboard equivalent. You can also use this command to switch into and out of italics as you are typing a Text cell. Similarly, you can change the size of the font, its color, *etc.*

Quitting the Kernel

Sometimes you may want to start a fresh *Mathematica* session without exiting the program, restarting, and reopening the Notebook(s) you are working on. This might happen, for example, if some previously defined variables or functions are causing problems that you can't easily eradicate. Or you may simply want to reset the In/Out numbering. One way to take care of these problems is to quit and restart the *Mathematica kernel*. When you quit the kernel, the underlying *Mathematica* process will die, along with any variable or function definitions, but your Notebooks will remain on the screen. The next time you press SHIFT+ENTER, a new kernel will start up, and after a slight delay you can keep working as before. The In/Out numbering will be initialized to 1.

To quit the kernel, you can select `Kernel:Quit Kernel:Local`. (In networked configurations, there may be other choices besides `Local`; you will then have to figure out which one is applicable in your situation.) A box will appear, asking you if you really want to quit the kernel; click on the appropriate button to confirm that you want to do so.

Evaluating the Notebook

If you save your work and resume it later in a new *Mathematica* session, none of the variables or functions you defined in the previous session will be in effect unless you reevaluate the Input cells that contain them. You can do this for each cell by clicking in the cell and pressing SHIFT+ENTER, or you can select `Kernel:Evaluation:Evaluate Notebook` to automatically evaluate all input cells in order from the top of the Notebook to the bottom. You may have to wait a while for all the calculations to be redone. Note that the In/Out numbering will almost certainly be different from that in your previous sessions with the Notebook, so a statement like **%24** may no longer refer to the appropriate output.

Customization and Printing

There are several ways to customize *Mathematica* Notebooks and sessions. You can change the way Notebooks are printed using the items in the `File:Printing Settings` menu. For example, you can adjust the margins by selecting `Printing Options....` Select `Headers and Footers...` to change what is printed at the top and bottom of each page. And in *Windows 95*, you can set options specific to your printer by selecting `Page Setup...`, or by clicking on "Properties" in the `Print...` dialog box.

To change how your Notebook looks overall, you use a *style sheet*. You can either select a predefined style sheet from the `Format:Style Sheet` menu, or you can create your own style sheet using `Format:Edit Style Sheet....` To learn more, choose the Other Information section of the Help Browser and click on "Menu Commands", then "Format Menu", then "Edit Style Sheet...".

Finally, you can change a large number of advanced formatting and printing options through the *Option Inspector*. To open the Option Inspector, select `Format:Option Inspector...` or `Edit:Preferences....` For instructions on how to use it, choose the Other Information section of the Help Browser and click on "Menu Commands", then "Format Menu", then "Option Inspector".

Using the Notebook Effectively

In preparing solutions to the problem sets in this book, you will use *Mathematica* to analyze various differential equations. You should make use of the editing features of the *Mathematica* Notebook interface to produce a polished document. In particular, material that is not relevant to the final answer (typing errors, trial calculations, *etc.*) should be deleted from the final printed form of the Notebook. Answers to interpretive questions should appear in a logical place relative to the *Mathematica* output.

You should take special care to make sure that Input cells are reevaluated after they are edited. Failure to do this can result in Output cells that are incompatible with their associated Input cells. This is confusing and misleading to anyone reading the Notebook.

Chapter 5

Solutions of
Differential Equations

In this chapter, we show how to solve differential equations with *Mathematica*. For many differential equations, the *Mathematica* command **DSolve** produces the general solution to the differential equation, or the specific solution to an associated initial value problem. We also discuss the existence, uniqueness, and stability of solutions of differential equations. These are fundamental issues in the theory and application of differential equations. An understanding of them helps in interpreting and using results produced by *Mathematica*.

Finding Symbolic Solutions

Consider the differential equation

$$\frac{dy}{dx} = f(x, y). \tag{1}$$

A solution to this equation is a function $y(x)$ of the independent variable x that satisfies $y'(x) = f(x, y(x))$. It is sometimes possible to find a formula for the solutions to (1); we call such a formula a *symbolic solution*. In *Mathematica*, the command that finds symbolic solutions is **DSolve**. To find a symbolic solution of the differential equation (1), type

```
DSolve[y'[x] == f[x, y[x]], y[x], x]
```

Notice that you must type **y[x]** instead of **y**, and that the equation is typed with a double equal sign. *Mathematica* produces the answer in terms of an arbitrary constant C[1]. (For higher order equations, there will be several arbitrary constants.) For example, consider the differential equation

$$\frac{dy}{dx} = x + y.$$

You can solve this equation in *Mathematica* by typing

```
DSolve[y'[x] == x + y[x], y[x], x]
```

The output from this command is

$$\{\{y[x] \to -1 - x + E^x C[1]\}\}$$

The solution of the differential equation is the expression following the arrow.

You can then obtain specific solutions by choosing specific values for C[1]. For example, the solution satisfying a given initial condition can be obtained by choosing C[1] appropriately. This value of C[1] can be found by imposing the initial condition on the general solution and solving for C[1]. Alternatively, you can specify the initial condition as well as the differential equation when you use **DSolve**. To solve the initial value problem

$$\frac{dy}{dx} = f(x, y), \qquad y(x_0) = y_0,$$

you would type

```
DSolve[{y'[x] == f[x, y[x]], y[x0] == y0}, y[x], x]
```

For example, the command

```
sol1 = DSolve[{y'[x] == x + y[x], y[0] == 2}, y[x], x]
```

produces the output

$$\{\{y[x] \to -1 + 3E^x - x\}\}$$

In this example, we have given the name **sol1** to the output.

Next, suppose you want to plot the solution, or find its value at a particular value of x. You can't just tell *Mathematica* to plot **y[x]** or evaluate **y[x]** at $x = 5$, say, because **DSolve** presents the solution in the form of a *transformation rule*, not a function. The most straightforward way to plot or evaluate the solution is to define a function equal to the expression following the arrow,

```
y1[x_] = -1 + 3 Exp[x] - x
```

and then plot or evaluate **y1[x]**. You can also use the *Mathematica* replacement operator **/.** to extract the function automatically from **sol1**. The command

```
y1[x_] = y[x] /. First[sol1]
```

does the job. This command is equivalent to the previous explicit definition of **y1[x]**. Transformation rules, the replacement operator, and the **First** command are explained in Chapter 8.

We often want to study a family of solutions obtained by varying the initial condition. Here is a natural way to do this in *Mathematica*. Begin by solving the differential equation with a generic initial value. For example,

```
sol2 = DSolve[{y'[x] == x + y[x], y[0] == c}, y[x], x]
```

produces a solution formula in terms of the initial value c. The result of this command is

$$\{\{y[x] \to -1 + (1 + c)E^x - x\}\}$$

Next, view the right-hand side as a function of both x and c. To define such a function in *Mathematica*, type

```
y2[x_, c_] = y[x] /. First[sol2]
```

Now suppose we want to plot a family of solution curves with initial values $y(0) = -2, -1, \ldots, 4$ on the interval $0 \leq x \leq 3$. We can type

```
Plot[Evaluate[Table[y2[x, c], {c, -2, 4}]], {x, 0, 3}]
```

The result of this command is shown in Figure 1. The **Evaluate** command is explained in Chapter 8; the **Table** command, in Chapter 3.

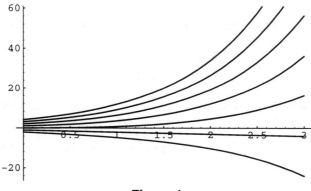

Figure 1

DSolve is a robust command that can solve many differential equations. In fact, it can solve most of the differential equations that can be solved with the standard solution methods one learns in an introductory course. However, there are many other equations, some of which can be solved by more advanced solution methods, that *Mathematica* cannot solve.

REMARK. Although we have focused on first order equations in this section, **DSolve** also solves higher order equations and systems of equations. For examples, see Chapter 8, Chapter 12, and the Sample Notebook Solutions.

Existence and Uniqueness

The fundamental existence and uniqueness theorems for differential equations guarantee that every initial condition $y(x_0) = y_0$ leads to a unique solution near x_0, provided that the right-hand side of the differential equation (1) is a "nice" function (to be specific, provided that f and $\partial f / \partial y$ are continuous functions). Graphically, these theorems say that there is a solution curve through every point and that the solution curves cannot cross. Thus, initial value problems (IVPs) have exactly one solution, but, since there are an infinite number of initial conditions, differential equations have an infinite number of solutions. This principle is implicit in the results obtained above with **DSolve**; when we do not specify an initial condition, the solution depends on an arbitrary constant; when we specify an initial condition, the solution is completely determined.

It is important to remember that the existence and uniqueness theorems only guarantee the existence of a solution *near* the initial point x_0. Consider the initial value problem

$$\frac{dy}{dx} = y^2, \qquad y(0) = 1.$$

From the *Mathematica* command

```
sol3 = DSolve[{y'[x] == y[x]^2, y[0] == 1}, y[x], x]
```

we get the output

$$\{\{y[x] \to \frac{1}{1-x}\}\}$$

To understand where the solution exists, we can graph the expression given by **DSolve** (see Figure 2 below).

```
Plot[y[x] /. sol3, {x, -1, 2}, PlotRange -> {-20, 20}];
```

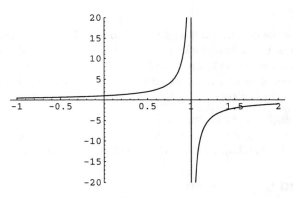

Figure 2

We see that the graph has two branches. Since the left branch passes through the initial data point $(0, 1)$, it is the desired solution. Furthermore, we see that the solution exists from $-\infty$ to 1, but does not extend beyond 1. In fact, it becomes unbounded as x approaches 1 from the left. The right branch of the graph depicts more of the function $\frac{1}{1-x}$, but it is not part of the solution of the IVP. Note that *Mathematica* includes the asymptote in its graph.

Stability of Differential Equations

In addition to existence and uniqueness, the sensitivity of the solution of an initial value problem to the initial value is a fundamental issue in the theory and application of differential equations. We introduce this issue through the following examples.

EXAMPLE 1. Consider the initial value problem

$$\frac{dy}{dx} + 2y = e^{-x}, \qquad y(0) = y_0.$$

We ask: How does the solution $y(x)$ depend on the initial value y_0? More specifically: Does the solution depend continuously on y_0? Do small variations in y_0 lead to small, or large, variations in the solution? Since the solution is

$$y(x) = e^{-x} + (y_0 - 1)e^{-2x},$$

we immediately see that, for any fixed x, the solution $y(x)$ depends continuously on y_0. We can say more. If we let $\tilde{y}(x)$ be the solution of the same equation, but with the initial condition $\tilde{y}(0) = \tilde{y}_0$, then $\tilde{y}(x) = e^{-x} + (\tilde{y}_0 - 1)e^{-2x}$. So, we have

$$|y(x) - \tilde{y}(x)| = |y_0 - \tilde{y}_0|e^{-2x}.$$

Thus for $x \geq 0$ we see that $|y(x) - \tilde{y}(x)|$ is never larger than $|y_0 - \tilde{y}_0|$. In fact, $|y(x) - \tilde{y}(x)|$ decreases as x increases. For $x \leq 0$ the situation is different. When x is a large negative number, the initial difference $|y_0 - \tilde{y}_0|$ is magnified by the large factor e^{-2x}. Even though $y(x)$ depends continuously on y_0, small changes in y_0 lead to large changes in $y(x)$. For example, if $|y_0 - \tilde{y}_0| = 10^{-3}$, then $|y(-7) - \tilde{y}(-7)| = 10^{-3}e^{14} \approx 1203$. These observations are confirmed by Figure 3, a plot of the solutions corresponding to initial values $y(0) = 0.97, 1, 1.03$ for $-3 \leq x \leq 3$.

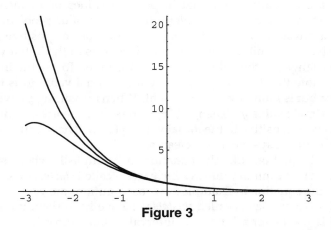

Figure 3

EXAMPLE 2. The solution of the initial value problem

$$\frac{dy}{dx} - 2y = -3e^{-x}, \qquad y(0) = y_0$$

is $y(x) = e^{-x} + (y_0 - 1)e^{2x}$. Again, $y(x)$ depends continuously on y_0 for fixed x. Letting $\tilde{y}(x)$ be the solution with initial value $\tilde{y}_0$, we see that

$$|y(x) - \tilde{y}(x)| = |y_0 - \tilde{y}_0|e^{2x}.$$

Now we see that $y(x)$ is very sensitive to changes in the initial value for x large and positive, but insensitive for x negative. These observations are confirmed by Figure 4, a plot of the solutions corresponding to initial values $y(0) = 0.97, 1, 1.03$ for $-2 \le x \le 2$.

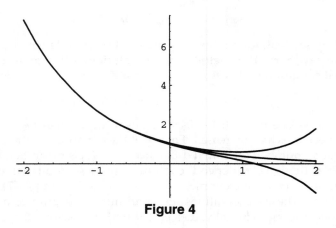

Figure 4

Often, we are primarily interested in positive values of x, such as when x corresponds to time in a physical problem. If an initial value problem is to predict the future of a physical system effectively, the solution for positive x should be fairly insensitive to the initial value; *i.e.*, small changes in the initial value should lead to small changes in the solution for positive time. To see the importance of this principle, note that for a physical system the initial value y_0 is typically not known exactly, but is found by measurement. When measuring y_0, we usually get only an approximate value $\tilde{y}_0$. Then, if $\tilde{y}(x)$ is the solution corresponding to $\tilde{y}_0$ and the solution is very sensitive to the initial value, $\tilde{y}(x)$ will have little relation to the actual state $y(x)$ of the system as x increases.

A differential equation (like the equation in Example 1) whose solutions are fairly insensitive to the initial value as x increases is called *stable*, while a differential equation (like the equation in Example 2) whose solutions are very sensitive to the initial value as x increases is called *unstable*. As we have seen, if an equation is unstable and is used over a long time interval, a small error in the initial value can result in a large error later on. Caution should be exercised when using an unstable equation to model a physical problem.

REMARK. We have considered stability as x increases; *i.e.*, stability to the right. We can also consider stability to the left. There are equations that are stable both to the left and to the right, and equations that are unstable both to the left and right.

The following theorem, which we state without proof, is often useful in assessing stability.

THEOREM. *Suppose $f(x, y)$ has continuous first order partial derivatives in the vertical strip*

$$R = \{(x, y) \ : \ x_0 \le x \le x_1, -\infty < y < \infty\},$$

and suppose there are numbers K and L such that

$$K \le \frac{\partial f}{\partial y}(x, y) \le L, \quad \text{for all } (x, y) \in R.$$

If $y(x)$ and $\tilde{y}(x)$ are solutions of $dy/dx = f(x, y)$ on the interval $x_0 \le x \le x_1$ with initial values $y(x_0) = y_0$ and $\tilde{y}(x_0) = \tilde{y}_0$, respectively, then

$$|y_0 - \tilde{y}_0|e^{K(x-x_0)} \le |y(x) - \tilde{y}(x)| \le |y_0 - \tilde{y}_0|e^{L(x-x_0)},$$

for all $x_0 \le x \le x_1$.

If $L \le 0$, then the right-hand inequality in the theorem shows that

$$|y(x) - \tilde{y}(x)| \le |y_0 - \tilde{y}_0|, \quad \text{for all } x_0 \le x \le x_1.$$

Thus the solutions differ by no more than the difference in the initial values, and the differential equation is stable. Moreover, if $L > 0$, but not too large, and $x_1 - x_0$ is not too large, then

$$|y(x) - \tilde{y}(x)| \le M|y_0 - \tilde{y}_0|, \quad \text{for all } x_0 \le x \le x_1,$$

where $M = e^{L(x_1 - x_0)}$ is a moderate-sized constant. Thus the equation is only mildly sensitive to changes in the initial value, and the equation is only mildly unstable. On the other hand, if $K > 0$, then the left-hand inequality in the theorem shows that the solution is sensitive to changes in the initial value, especially over long intervals. We can briefly summarize these results by saying that if $\partial f/\partial y \le 0$, then the differential equation is stable; but if $\partial f/\partial y > 0$, then the equation is unstable.

REMARK. The right-hand inequality in the theorem is an example of a *continuous dependence* result; it shows that the solution depends continuously on the initial value.

Let us examine our examples in light of these observations. Rewriting the equation in Example 1 as $dy/dx = -2y + e^{-x}$, we see that $f(x, y) = -2y + e^{-x}$ and $\partial f/\partial y = -2$. We can apply the theorem with $x_0 = 0$, $x_1 = \infty$, and $L = K = -2$ to find that

$$|y(x) - \tilde{y}(x)| = |y_0 - \tilde{y}_0|e^{-2x}, \quad \text{for all } x \ge 0,$$

as we found above from the solution formula. Thus the equation is stable. We could also conclude that the equation is stable just by noting that $\partial f/\partial y < 0$. Likewise, for the equation of Example 2, $f(x, y) = 2y - 3e^{-x}$ and $\partial f/\partial y = 2 > 0$, so the equation is unstable.

Of course, we can understand the stability of the differential equations in these two examples by examining the solution formulas. But $\partial f/\partial y$ can be calculated and its sign and size found, even if a solution formula cannot be found. For example, we can immediately tell that the differential equation $dy/dx + x^2 y^3 = \cos x$ is stable because

$$f(x, y) = -x^2 y^3 + \cos x$$

and $\partial f/\partial y = -3x^2 y^2 \leq 0$. Yet neither **DSolve** nor any other standard technique enables us to find a formula solution to this differential equation.

Finally, we note that many equations of the form $dy/dx = f(x, y)$ cannot be classified simply as stable or unstable, because $\partial f/\partial y$ may be negative at some points and positive at others. Nonetheless, we may still be able to determine whether a particular solution is stable (insensitive to its initial value) or unstable (sensitive to its initial value) according to whether $\partial f/\partial y \leq 0$ or $\partial f/\partial y > 0$ along the solution curve. Throughout the book we will see many examples that illustrate the dependence (either sensitive or insensitive) of the solution to an initial value problem on the initial value.

CAVEAT. It is important to realize that *Mathematica*, like any software system, may occasionally produce misleading or incorrect results. Consider, for example, the initial value problem

$$\frac{dy}{dx} = xy^2, \qquad y(0) = 0.$$

In Version 3.0 of *Mathematica*, the command

```
DSolve[{y'[x] == x*y[x]^2, y[0] == 0}, y[x], x]
```

gives the output

```
        Power::infy : Infinite expression 1/0 encountered.
        { }
```

suggesting that there is no solution. In fact, the existence and uniqueness theorem applies to this IVP, and the function $y(x) \equiv 0$ is the unique solution. This example shows that a good theoretical understanding is a valuable guide in interpreting computer-generated results, and therefore in identifying situations in which *Mathematica* has produced a misleading or incorrect result. Incidentally, the reader may check that *Mathematica* can find a general solution of $y' = xy^2$, and the unique solution of the IVP $y' = xy$, $y(0) = 0$.

Chapter 6

A Qualitative Approach
to Differential Equations

In this chapter, we discuss a qualitative approach to the study of differential equations. With this approach, we obtain qualitative information about the solutions directly from the differential equation, without the use of a solution formula.

Consider the general first order differential equation

$$\frac{dx}{dt} = f(t, x). \tag{1}$$

We can obtain qualitative information about the solutions $x(t)$ by viewing (1) geometrically. Specifically, we can obtain this information from the direction field of (1). Recall that the direction field is obtained by drawing through each point in the (t, x)-plane a short line segment with slope $f(t, x)$. Solutions, or integral curves, of (1) have the property that at each of their points they are tangent to the direction field at that point, and therefore the general *qualitative* nature of the solutions can be determined from the direction field. Direction fields can be drawn by hand for some simple differential equations, but *Mathematica* can draw them for any first order equation. We illustrate the qualitative approach with two examples.

EXAMPLE 1. Consider the equation

$$\frac{dx}{dt} = e^{-t} - 2x. \tag{2}$$

Mathematica's command for plotting direction fields is **PlotVectorField**, in the **PlotField** package. To load this package, type <<**Graphics`PlotField`** (note that these are backward single quotes). See the *Packages* section of Chapter 3 for important information on using packages. You can then plot the direction field of (2) on the rectangle $-2 \leq t \leq 3$, $-1 \leq x \leq 2$ by typing

```
PlotVectorField[{1, Exp[-t] - 2x}, {t, -2, 3}, {x, -1, 2},
    ScaleFunction -> (1&), Axes -> True, Ticks -> None,
    Frame -> True, AspectRatio -> 1]
```

The resulting direction field is shown in Figure 1. The **ScaleFunction** option ensures that the arrows have equal length. It and the **AspectRatio** option are discussed in the *Graphics* section in Chapter 8; see also the online help. Also, the **1** in `{1, Exp[-t] - 2x}` is present because **PlotVectorField** normally plots the vector field of a system of two differential equations, and we only have a single equation here. (See the *Remarks* at the end of Chapter 13.)

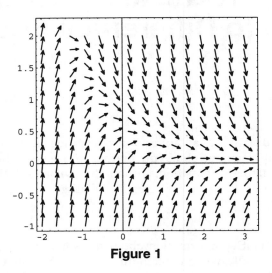

Figure 1

The direction field strongly suggests that if a solution is negative at some point, then it is increasing at that point, and that all solutions approach zero as $t \to \infty$. The general solution of equation (2) is $e^{-t} + ce^{-2t}$.

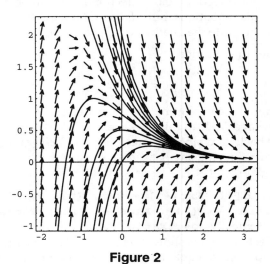

Figure 2

Figure 2 shows the direction field together with several of these solutions. We can also deduce from the differential equation itself that the maximum points on the solution curves depicted in Figure 2 will occur along the curve $2x = e^{-t}$, since that is where $x' = 0$.

EXERCISE. Pursue this idea further by differentiating (2) and showing that the inflection points on the solution curves lie along the curve $4x = 3e^{-t}$.

Equation (2) can be solved explicitly because it is linear. Now let us consider an example that cannot be solved explicitly by any of the standard solution methods.

EXAMPLE 2. Consider the equation

$$\frac{dx}{dt} = x^2 + t. \tag{3}$$

Its direction field is obtained with the *Mathematica* command

```
PlotVectorField[{1, x^2 + t}, {t, -2, 2}, {x, -2, 2},
  Axes -> True, ScaleFunction -> (1&),
  Ticks -> None, Frame -> True]
```

and is shown in Figure 3.

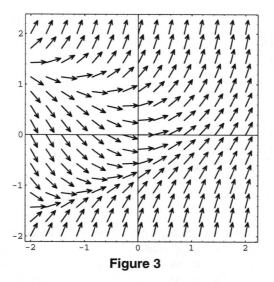

Figure 3

From the direction field, it appears that all solutions are increasing for $t > 0$, and that all solutions eventually become positive. Figure 3 also suggests that all solutions approach infinity. Does this happen at a finite value of t or only as $t \to \infty$? It is imposible to decide on the basis of Figure 3, but in fact for each solution $x(t)$ there is a finite value t^* such that $\lim_{t \to t^*} x(t) = \infty$. Later, we will again encounter equations that cannot be solved explicitly by formula techniques. We will apply a combination of qualitative and numerical tools to analyze such equations, and in particular establish the fact just asserted about the solutions to equation (3) approaching infinity in finite time.

EXERCISE. Use **PlotVectorField** to graph some of the direction fields in your textbook.

Autonomous Equations

Equations of the form

$$\frac{dx}{dt} = f(x), \tag{4}$$

which do not involve t in the right-hand side, are called *autonomous* equations. Autonomous equations represent physical systems whose rules of evolution do not change with time and are particularly susceptible to qualitative analysis.

Consider equation (4); to be concrete, suppose $f(x)$ has two zeros x_1 and x_2, called *critical points* of the differential equation. Furthermore, suppose the graph of $f(x)$ is as shown in Figure 4.

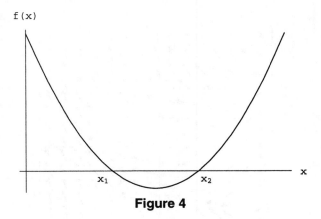

Figure 4

Then by considering the properties of $f(x)$, the direction field of (4) can be easily understood and drawn by hand. At the points $(0, x_1)$ and $(0, x_2)$ the slope is zero; at points $(0, x)$ with $x < x_1$, the slope is positive and increases from 0 to ∞ as x decreases from x_1 to $-\infty$; at points $(0, x)$ with $x_1 < x < x_2$, the slope is negative, and first decreases and then increases as x increases from x_1 to x_2; at points $(0, x)$ with $x > x_2$ the slope is positive and increases from 0 to ∞ as x increases from x_2 to ∞; finally, the slopes along any horizontal line are constant (since f does not depend on t). The direction field thus has the general appearance shown in Figure 5. Note that Figure 4 depicts $f(x)$ vs. x, while Figure 5 shows x vs. t.

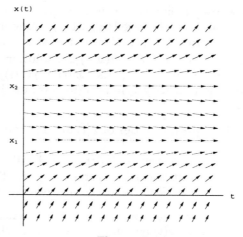

x(t)

x_2

x_1

t

Figure 5

Several facts are suggested by the direction field, namely, that: (i) there are two constant solutions $x = x_1$ and $x = x_2$; (ii) solutions starting above x_2 tend to ∞; and (iii) all other solutions tend to x_1 as $t \to \infty$. In fact, we can derive these properties of the solutions $x(t)$ directly from the graph of $f(x)$ as follows.

(a) The constant functions $x(t) = x_1$ and $x(t) = x_2$ are solutions, called the *equilibrium solutions*. They are the unique solutions that satisfy the initial conditions $x(0) = x_1$ and $x(0) = x_2$, respectively.

(b) Consider a solution $x(t)$ with $x(0) < x_1$. Then, because of the uniqueness of solutions, it must be that $x(t) < x_1$ for all t and

$$x'(t) = f(x(t)) > 0.$$

Hence $x(t)$ is an increasing function. Since it is also bounded, it has a limit at infinity

$$\lim_{t \to \infty} x(t) = b \leq x_1. \tag{5}$$

Could it happen that $b < x_1$? The answer is no, because if $b < x_1$, then

$$x'(t) = f(x(t)) \to f(b) > 0. \tag{6}$$

But equations (5) and (6) say that $x(t)$ approaches a horizontal asymptote at the same time that its slope is "permanently" bigger than a positive number. The resulting contradiction guarantees that

$$\lim_{t \to \infty} x(t) = x_1.$$

(c) If $x(t)$ is a solution with $x(0) > x_2$, then $x(t) > x_2$ for all t. Also

$$x'(t) = f(x(t)) > 0.$$

Hence $x(t)$ is once again increasing. In fact $x(t) \to \infty$. This can be shown by an argument similar to that used in part (b). Sometimes, as we shall see from the example below, $x(t)$ actually reaches ∞ in finite time.

(d) Now consider $x_1 < x_0 < x_2$. Then if $x(t)$ is a solution with $x(0) = x_0$, we must have $x_1 < x(t) < x_2$ for all t. Also, $x'(t) = f(x(t)) < 0$. Hence $x(t)$ is a decreasing function. Exactly as in part (b), it is not difficult to show that $\lim_{t \to \infty} x(t) = x_1$.

Note that in this analysis the properties of $x(t)$ are determined solely from the sign of $f(x)$.

We call the critical point x_1 *asymptotically stable* and call the critical point x_2 *unstable*, since solutions that start near x_1 converge to x_1, and those that start near x_2 not only do not converge to x_2, but eventually move away from x_2.

EXERCISE. Let $\bar{x}$ be the value between x_1 and x_2 where f takes on its minimum. Show that if $\bar{x} < x(0) < x_2$, then $x(t)$ has an inflection point where $x(t) = \bar{x}$.

Now we examine a specific example.

EXAMPLE 3. Consider the equation

$$\frac{dx}{dt} = x^2 - x.$$

Its direction field has the general appearance of Figure 5, with $x_1 = 0$ and $x_2 = 1$. But we can actually derive an explicit formula solution. If $x \neq 0$ and $x \neq 1$, we can solve by separating variables:

$$
\begin{aligned}
t + C = \int dt &= \int \frac{dx}{x^2 - x} \\
&= \int \left(\frac{1}{x - 1} - \frac{1}{x} \right) dx \\
&= \ln |x - 1| - \ln |x| \\
&= \ln \left| \frac{x - 1}{x} \right| \\
&= \ln \left| 1 - \frac{1}{x} \right|.
\end{aligned}
$$

Exponentiating both sides, we obtain

$$1 - \frac{1}{x} = ke^t,$$

where $k = \pm e^C$. Solving for x, we find that $x(t) = 1/(1 - ke^t)$. Noting that $x_0 = x(0) = 1/(1 - k)$ we have

$$x(t) = \frac{x_0}{(1 - x_0)e^t + x_0}. \tag{7}$$

Although formula (7) was derived under the assumption that $x_0 \neq 0, 1$, it is easily seen to be valid for all values of x_0.

Mathematica can solve this equation and plot the solution curves. Here is a sequence of commands that does so.

```
xsol[t_, c_] = x[t] /. First[DSolve[
  {x'[t] == x[t]^2 - x[t], x[0] == c}, x[t], t]];
Plot[Evaluate[Table[xsol[t, c], {c, -1.6, 1.6, 0.2}]],
  {t, 0, 3}, Ticks -> None, AxesLabel -> {"t", "x(t)"}];
```

Figure 6 contains the actual solution curves for the differential equation—as drawn by *Mathematica*. Note that the solution curves corresponding to $x(0) > 1$ go to infinity in finite time, and *Mathematica* includes the vertical asymptotes.

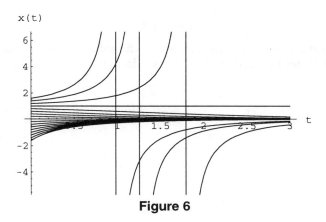

Figure 6

EXERCISES.

(a) Use formula (7) to verify the properties of $x(t)$ obtained above by the qualitative method. Determine where the solutions are increasing or decreasing, and whether there are limits as $t \to \infty$.

(b) Suppose $x_0 > 1$. Use formula (7) to show that $x(t) \to \infty$ in finite time. Find the time t^* at which this happens.

(c) Show that $x(t) = 0$ and $x(t) = 1$ are the equilibrium solutions. Does formula (7) yield these solutions?

Here is another example. The left graph in Figure 7 shows a function $f(x)$ with three zeros; thus, the autonomous differential equation $x' = f(x)$ has three critical points. Let's look at the critical point in the middle. Solutions starting below b have a positive slope, so they increase toward b. Solutions starting above b have a negative slope, so they decrease toward b. Therefore, we expect b to be an asymptotically stable equilibrium solution. A similar analysis of the signs of $f(x)$ suggests that the other two critical points are unstable.

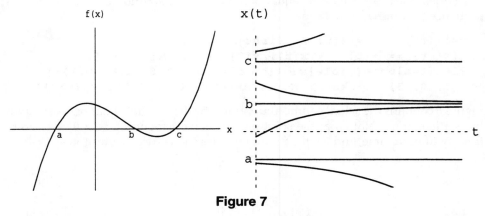

Figure 7

The right graph in Figure 7 shows a few of the solution curves. As you can see, this graph illustrates the conclusions that we drew from our analysis of the graph of $f(x)$.

Problem Set B

First Order Equations

The solutions to Problems 1 and 4 appear in the *Sample Notebook Solutions*.

1. The initial value problem

$$xy' + 2y = \sin x, \qquad y(\pi/2) = 1$$

has the solution

$$y(x) = x^{-2} \left(\frac{\pi^2}{4} - 1 - x \cos x + \sin x \right).$$

(a) Verify this using *Mathematica*. Then define the function $y(x)$ in *Mathematica*.

(b) Make a list of the values of $y(x)$ at $x = 0.5, 1, 1.5, 2, \ldots, 5$. (This can be done using **Table** and **TableForm** (see Chapter 3). Remember to use the **N** command to get numerical values.)

(c) Graph $y(x)$ on the intervals $0 < x \leq 2$, $1 \leq x \leq 10$, and $10 \leq x \leq 100$. Describe the behavior of the solution near $x = 0$ and for large values of x.

(d) Find the solutions $y_j(x)$ of the differential equation corresponding to the initial conditions

$$y_j(\pi/2) = 0.2j, \qquad j = 1, \ldots, 5.$$

Plot the functions $y_j(x)$, $j = 1, \ldots, 5$, on the same graph.

(e) What do the solutions have in common near $x = 0$? for large values of x? Can you find a solution to the differential equation that has no singularity at $x = 0$? If so, graph it.

2. Consider the initial value problem

$$xy' + y = 2x, \qquad y(1) = c.$$

(a) Solve it using *Mathematica*.

(b) Use *Mathematica* to graph the solutions for $c = 0.8, 0.9, 1, 1.1, 1.2$ on the interval $(0.75, 1.25)$.

(c) Evaluate these solutions at $x = 0.01, 0.1, 1, 10$.

(d) Now plot all five solutions together on the interval $(0, 2.5)$. How do changes in the initial data affect the solution as $x \to \infty$? as $x \to 0^+$?

(See the section *Lists and Tables* in Chapter 3.)

3. Solve the initial value problem

$$y' - y = \cos x, \qquad y(0) = c.$$

Use *Mathematica* to graph solutions for $c = -0.9, -0.8, \ldots, -0.1, 0$. Display all the solutions on the same interval between $x = 0$ and an appropriately chosen right endpoint. Explain what happens to the solution curves for large values of x. (*Hint:* You should identify three distinct types of behavior.) Now, based on this problem, and the material in Chapters 5 and 6, discuss what effect small changes in initial data can have on the global behavior of solution curves.

4. Consider the differential equation

$$\frac{dy}{dx} = \frac{x - e^{-x}}{y + e^y}$$

(*cf.* Problem 7, Section 2.3 in Boyce & DiPrima).

(a) Solve it using *Mathematica*. Observe that the solution is given implicitly in the form

$$f(x, y) = c.$$

(b) Use **ContourPlot** (*cf.* Problem Set A) to see what the solution curves look like. For your x and y ranges, you might use **{x,-1,3}** and **{y,-2,2}**.

(c) Plot the solution satisfying the initial condition $y(1.5) = 0.5$. Your plot should show two curves. Indicate which one corresponds to the solution.

(d) Given a value x_1, to find $y(x_1)$ you need to find the solution of

$$f(x_1, y) = f(1.5, 0.5)$$

(viewed as an equation in y) near $y = 0.5$. This can be done using **FindRoot**. Find $y(0), y(1), y(1.8), y(2.1)$. Mark these values on your plot.

5. Consider the differential equation

$$(e^x \sin y - 2y \sin x) + (e^x \cos y + 2 \cos x) \frac{dy}{dx} = 0,$$

(*cf.* Problem 7, Section 2.8 in Boyce & DiPrima).

(a) Solve it using *Mathematica*. Observe that the solution is given implicitly in the form

$$f(x, y) = c.$$

(*Hint*: Enter the equation exactly as is. *Mathematica* cannot handle it if you solve the equation for $y'(x)$ and enter it in that form.)

(b) Use **ContourPlot** to see what the solution curves look like. For your x and y ranges, you might use **{x,-3,3}** and **{y,-3,3}**.

(c) Use **ImplicitPlot** to plot the solution satisfying the initial condition $y(0) = 0.5$. Your plot may show several curves. Indicate which one corresponds to the solution.

(d) Given a value x_1, to find $y(x_1)$ you need to find the solution of

$$f(x_1, y) = f(0, 0.5)$$

(viewed as an equation in y) near $y = 0.5$. This can be done using **FindRoot**. Find $y(-1), y(1), y(2)$. Mark these values on your plot.

6. In this problem, we study continuous dependence of solutions on initial data.

   (a) Solve the initial value problem

   $$y' = y/(1 + x^2), \qquad y(0) = c.$$

   (b) Let y_c denote the solution in part (a). Use *Mathematica* to plot the solutions y_c for $c = -10, -9, \ldots, -1, 0, 1, \ldots, 10$ on one graph. Display all the solutions on the interval $-20 \le x \le 20$. You may need to employ the **PlotRange** option to ensure that all the graphs are fully displayed.

   (c) Compute $\lim_{x \to \pm\infty} y_1(x)$.

   (d) Now find a single constant L such that for all real x, we have

   $$|y_a(x) - y_b(x)| \le L|a - b|,$$

   for any pair of numbers a and b. Show that $L = 1$ will work if we consider only negative values of x.

7. Use **DSolve** to solve the following differential equations or initial value problems from Boyce & DiPrima. In some cases, *Mathematica* will not be able to solve the equation. (Before moving on to the next equation, make sure you haven't mistyped something.) In other cases, *Mathematica* may give extraneous solutions. (Sometimes these correspond to imaginary roots of the equation it solves to get y in terms of x.) If so, you should indicate which solution or solutions are valid. You also might try entering alternative forms of an equation, for example, $M + Ny' = 0$ instead of $y' = -M/N$, or vice versa.

   (a) $x^3 y' + 4x^2 y = e^{-x}, y(-1) = 0$ (Sect. 2.1, Prob. 19)

(b) $y' = ry - ky^2$ (Sect. 2.2, Prob. 39)

(c) $y' = x(x^2 + 1)/4y^3, y(0) = -1/\sqrt{2}$ (Sect. 2.3, Prob. 16)

(d) $(2x - y)\, dx + (2y - x)\, dy = 0, y(1) = 3$ (Sect. 2.8, Prob. 13)

(e) $dy/dx = (2y - x)/(2x - y)$ (Sect. 2.9, Prob. 15(a))

(f) $dy/dx = (2x + y)/(3 + 3y^2 - x), y(0) = 0$ (Sect. 2.10, Prob. 3).

8. Use **DSolve** to solve the following differential equations or initial value problems from Boyce & DiPrima. See Problem 7 for additional instructions.

(a) $xy' + (x + 1)y = x, y(\ln 2) = 1$ (Sect. 2.1, Prob. 20)

(b) $y' - (1/x)y = x^{1/2}$ (Sect. 2.2, Prob. 19)

(c) $y' = 2(1 + x)(1 + y^2), y(0) = 0$ (Sect. 2.3, Prob. 26)

(d) $(x \ln y + xy)\, dx + (y \ln x + xy)\, dy = 0; x > 0, y > 0$ (Sect. 2.8, Prob. 11)

(e) $(x^2 + 3xy + y^2)\, dx - x^2 dy = 0$ (Sect. 2.9, Prob. 8)

(f) $dy/dx = -(2xy + y^2 + 1)/(x^2 + 2xy)$ (Sect. 2.10, Prob. 5).

9. Chapter 6 describes how to plot the direction field for a first order equation. For each equation below, plot the direction field on a rectangle large enough (but not too large) to show clearly all of its equilibrium points. Find the equilibria and state whether each is stable or unstable. If you cannot determine the precise value of an equilibrium point from the direction field, use **FindRoot**.

(a) $y' = -y(y - 2)(y - 4)/10$

(b) $y' = y^2 - 3y + 1$

(c) $y' = 0.1y - \sin y$.

You may want to experiment with the plotting options **ScaleFunction**, **ScaleFactor**, or **AspectRatio** to improve your pictures.

10. In this problem, we use the direction field capabilities of *Mathematica* to study two nonlinear equations, one autonomous and one nonautonomous.

(a) Plot the direction field for the equation

$$\frac{dy}{dx} = 3\sin y + y - 2$$

on a rectangle large enough (but not too large) to show all possible limiting behaviors of solutions as $x \to \infty$. Find approximate values for all the equilibria of the system (you should be able to do this with **FindRoot** using guesses based on the direction field picture), and state whether each is stable or unstable.

(b) Plot the direction field for the equation

$$\frac{dy}{dx} = y^2 - xy,$$

again using a rectangle large enough to show the possible limiting behaviors. Identify the unique constant solution. Why is this solution evident from the differential equation? If a solution curve is ever below the constant solution, what must its limiting behavior be as $x \to +\infty$? For solutions lying above the constant solution, describe two possible limiting behaviors as $x \to +\infty$. There is a solution curve that lies along the boundary of the two limiting behaviors. Explain (from the differential equation) why that solution curve cannot be bounded. Can you guess how its limiting behavior might differ from that of the two limiting behaviors you have identified?

(c) Confirm your analysis by using **DSolve** on the system $y' = y^2 - xy$, $y(0) = c$, and then examining different values of c.

11. The solution of the differential equation

$$y' = \frac{2y - x}{2x - y}$$

is given implicitly by $|x - y| = c|x + y|^3$ (Boyce & DiPrima, Problem 15a, Section 2.9; you need not verify the solution with **DSolve**). However, it is difficult to understand the solutions directly from this algebraic information.

(a) Use **PlotVectorField** to plot the direction field of the differential equation. Note that *Mathematica* plots the direction field by computing vectors at the points of a regularly spaced grid. In this problem, if the grid contains a point on the line $y = 2x$ (where the denominator of (i) vanishes and the equation is singular), then an error message ensues. You can avoid this problem by choosing the endpoints of your rectangles so that the grid does not contain any points on the singular line. One way to achieve this is by selecting irrational endpoints, as in the following example:

```
PlotVectorField[{1, 1/(x + y)}, {x, -1, Sqrt[2]},
    {y, -1, Sqrt[2]}]
```

(b) Use **ContourPlot** to plot the solutions with initial conditions $y(2) = 1$ and $y(0) = -3$. (Note that **Abs** is the absolute value function in *Mathematica*.) Use **Show** to put these plots and the vector field plot together on the same graph.

(c) For the two different initial conditions in part (b), use your pictures to estimate the largest interval on which the unique solution function is defined.

12. Consider the differential equation

$$x' = -tx^2.$$

(a) Use *Mathematica* to plot the direction field of the differential equation. Is there a constant solution? If $x(0) > 0$, what happens as t increases? If $x(0) < 0$, what happens as t increases?

(b) Use **DSolve** to solve the differential equation, thereby obtaining a solution of the form

$$x(t) = \frac{2}{t^2 + 2c}$$

(where c stands for either C[1] or -C[1] depending on your version of *Mathematica*). Note that $x(0) = 1/c$. In graphing $x(t)$, it helps to consider three separate cases: (i) $c > 0$, (ii) $c = 0$, and (iii) $c < 0$. In each case, graph the solution for several specific values of c, and identify important features of the curves. In particular, in case (iii), compute the location of the asymptotes in terms of c.

(c) Identify five different types of solution curves for the differential equation. In each case, specify the t-interval of existence, whether the solution is increasing or decreasing, and the limiting (asymptotic) behavior at the ends of the interval.

(d) Use **Show** to combine the direction field plotted in (a) with the graphs plotted in (b). (*Hint*: You may have to use **PlotRange** to keep the scales compatible.)

13. Consider the following logistic-with-threshold model for population growth:

$$x' = x(1 - x)(x - 3).$$

(a) Find the equilibrium solutions of the differential equation. Now draw the direction field, and use it to decide which equilibrium solutions are stable and which are unstable. In particular, what is the limiting behavior of the solution if the initial population is between 1 and 2? greater than 3? exactly 1? between 1 and 2?

(b) Next replace the logistic law by the Gompertz model, but retain the threshold feature. The equation becomes

$$x' = x(1 - \ln x)(x - 3).$$

Once again, find the equilibrium solutions and draw the direction field. You may have trouble near $x = 0$ (because of the logarithm). Use a range like $0.01 \le x \le 4$ to circumvent that problem. But you will also have difficulty "reading the field" between 2.5 and 3. There appears to be a continuum of equilibrium solutions.

(c) Plot the function $f(x) = x(1 - \ln x)(x - 3)$ on the interval $0.01 \le x \le 4$, and then use **Limit** to evaluate $\lim_{x \to 0} f(x)$.

(d) Use these plots and the discussion in Chapter 6 to decide which equilibrium solutions are stable and which are unstable. Now use the last plot to explain why the direction field (for $2.5 \le x \le 3$) appears so inconclusive regarding the stability of the equilibrium solutions. (*Hint*: The maximum value of f is a relevant number. You can find it by applying **FindMinimum** to $-f$.)

14. This problem is based on Example 3 in Section 2.5 of Boyce & DiPrima: "A tank contains Q_0 lb of salt dissolved in 100 gal of water. Water containing $\frac{1}{4}$

lb of salt per gallon enters the tank at a rate of 3 gal/min, and the well-stirred solution leaves the tank at the same rate. Find an expression for the amount of salt $Q(t)$ in the tank at time t."

The differential equation

$$Q'(t) = 0.75 - 0.03Q(t)$$

models the problem (*cf.* equation (17) in Section 2.5).

(a) Plot the right-hand side of the differential equation as a function of Q, and identify the critical point.

(b) Analyze the long-term behavior of the solution curves by examining the sign of the right-hand side of the differential equation, in a similar fashion to the discussion in Chapter 6.

(c) Use *Mathematica* to plot the direction field of the differential equation. In choosing the rectangle for the direction field be sure to include the point $(0,0)$ and the critical value of Q.

(d) Use **DSolve** to find the solution $Q(t)$ and plot it for several specific values of Q_0. Do the solutions behave as indicated in parts (b) and (d)? You should combine the direction field plot from (c) with that of the solution curves.

15. A 10-gallon tank contains a mixture consisting of 1 gallon of water and an undetermined number $S(0)$ of pounds of salt in the solution. Water containing 1 lb/gal of salt begins flowing into the tank at the rate of 2 gal/min. The well-mixed solution flows out at a rate of 1 gal/min. Derive the differential equation for $S(t)$, the number of pounds of salt in the tank after t minutes, that models this physical situation. (*Note:* At time $t = 0$ there is 1 gallon of solution, but the volume increases with time.) Now draw the direction field of the differential equation on the rectangle $0 \le t \le 10$, $0 \le S \le 10$. From your plot,

(a) find the value A of $S(0)$ below which the amount of salt is a constantly increasing function, but above which the amount of salt will temporarily decrease before increasing;

(b) indicate how the nature of the solution function in case $S(0) = 1$ differs from all other solutions.

Now use **DSolve** to solve the differential equation. Reinforce your conclusions above by

(c) algebraically computing the value of A;

(d) giving the formula for the solution function when $S(0) = 1$;

(e) giving the amount of salt in the tank (in terms of $S(0)$) when it is at the point of overflowing;

(f) computing, for $S(0) > A$, the minimum amount of salt in the tank, and the time it occurs;

(g) explaining what principle guarantees the truth of the following statement: If two solutions S_1, S_2 correspond to initial data $S_1(0)$, $S_2(0)$ with $S_1(0) < S_2(0)$, then for any $t \geq 0$, it must be that $S_1(t) < S_2(t)$.

16. In this problem, we use **DSolve** and either **Solve** or **FindRoot** to model some population data. The procedure will be:

   (i) Assume a model differential equation involving unknown parameters.

   (ii) Use **DSolve** to find the solution of the differential equation in terms of the parameters.

   (iii) Use **Solve** or **FindRoot** to find the values of the parameters that fit the given data.

   (iv) Make predictions based on the results of the previous steps.

   (a) Let's use the model

$$\frac{dp}{dt} = ap + b, \quad p(0) = c,$$

where p represents the population at time t. Check to see that **DSolve** can solve this initial value problem in terms of the unknown constants a, b, c. Then define a function that expresses the solution at time t in terms of a, b, c, and t. Give physical interpretations to the constants a, b and c.

   (b) Next, let's try to model the population of Nevada, currently the fastest growing state in the U.S. Here is a table of census data:

Year	Population in thousands
1950	160.1
1960	285.3
1970	488.7
1980	800.5
1990	1201.8

We would like to find the values of a, b, and c that fit the data. However, with three unknown constants we will not be able to fit five data points. Use **Solve** or **FindRoot** to find the values of a, b, and c that give the correct population for the years 1960, 1970, and 1980. We will later use the data from 1950 and 1990 to check the accuracy of the model. *Important:* In this part let t represent the time in years since 1960, because *Mathematica* may get stuck if you ask it to fit the data at such high values of t as 1960–1980. You may have to experiment with your initial guesses for a, b, and c, and/or increase the option **MaxIterations** beyond its default value of 15 in order to get **FindRoot** to work.

   (c) Now define a function of t that expresses the predicted population in year t using the values of a, b, c found in part (b). Find the population this model gives for 1950 and 1990, and compare with the values in the table above. Use the model to predict the population of Nevada in the year 2000, and to predict when the population will reach 3 million. How would you adjust these predictions based on the 1950 and 1990 data? What adjustment to a and/or b might you make? Finally, graph the population the model gives from 1950 to 2050, and describe the predicted future of the population of Nevada, including the limiting population (if any) as $t \to \infty$.

17. Consider $x' = (\alpha - 1)x - x^3$.

(a) Use **Solve** to find the roots of $(\alpha - 1)x - x^3$. Explain why $x = 0$ is the only real root when $\alpha \leq 1$, and why there are three distinct real roots when $\alpha > 1$.

(b) For $\alpha = -2, -1, 0$, draw a direction field for the differential equation and deduce that there is only one equilibrium solution. What is it? Is it stable?

(c) Do the same for $\alpha = 1$.

(d) For $\alpha = 1.5, 2$, draw the direction field. Identify all equilibrium solutions, and describe their stability.

(e) Explain the following statement: "As α increases through 1, the stable solution $x = 0$ *bifurcates* into two stable solutions."

Chapter 7

Numerical Methods

Even though many differential equations can be solved explicitly in terms of the elementary functions of calculus, many others cannot. Consider, for example, the equation

$$\frac{dy}{dx} = e^{-x^2}.$$

Its solutions are the integrals, or antiderivatives, of e^{-x^2},

$$y(x) = \int e^{-x^2} \, dx + C,$$

but it is known that these integrals cannot be expressed in terms of the elementary functions of calculus. Similarly, the solution to the associated initial value problem

$$\frac{dy}{dx} = e^{-x^2}, \qquad y(x_0) = y_0$$

can be written as

$$y(x) = y_0 + \int_{x_0}^{x} e^{-s^2} \, ds,$$

but this formula cannot be simplified further; specifically, it is not an elementary function. (An *elementary function* is defined to be a polynomial, power, trigonometric, inverse trigonometric, exponential, or logarithmic function, or any combination of them via the processes of algebra and of composition and inversion.) More dramatic is the "simple" equation

$$\frac{dy}{dx} = x + y^2.$$

The solutions to this equation cannot be written in terms of elementary functions, or even in terms of repeated integrals of elementary functions!

When we cannot find an antiderivative in terms of elementary functions, we turn to a numerical method such as the trapezoid rule or Simpson's rule to evaluate definite integrals. Similarly, when we cannot solve a differential equation explicitly, or if the formula we find is too complicated, we also turn to numerical methods to solve the initial value problem.

Numerical Solutions Using *Mathematica*

Suppose we are interested in finding the solution to the initial value problem

$$\frac{dy}{dx} = f(x, y), \qquad y(x_0) = y_0$$

on an interval $a \leq x \leq b$ containing x_0, and suppose that we do not have a formula for $y(x)$. In such a situation, our strategy will be to produce a function $y_a(x)$ that both is a good approximation to $y(x)$ and can be calculated for any $x \in [a, b]$. The subscript a on y_a stands for *approximate solution*. Such a function $y_a(x)$ can be found with **NDSolve**, *Mathematica*'s numerical differential equation solver. We illustrate the use of **NDSolve** by considering the initial value problem

$$\frac{dy}{dx} = \frac{x}{y}, \qquad y(0) = 1. \tag{1}$$

Its exact solution is easily found to be

$$y(x) = \sqrt{x^2 + 1}.$$

Since we have an explicit formula for this solution, we will be able to compare the approximate solution $y_a(x)$ and the exact solution $y(x)$. To obtain $y_a(x)$ on the interval $[0, 2]$, type

```
ivp1 = {y'[x] == x/y[x], y[0] == 1}
sol1 = NDSolve[ivp1, y[x], {x, 0, 2}]
```

The first argument to **NDSolve** is a list that specifies the differential equation and the initial condition. The second argument specifies the dependent variable **y[x]**. The third argument, **{x, 0, 2}**, is a list consisting of the independent variable and the endpoints of the interval on which we wish to calculate the approximate solution.

The preceding commands produce the output

$$\{\{y[x] \rightarrow \text{InterpolatingFunction}[\{0., 2.\}, <>][x]\}\}$$

This is not quite the desired form of the approximate solution. To get the desired form, we use the replacement operator and the **First** command (exactly as when we introduced **DSolve** in Chapter 5). We type

```
ya1[x_] = y[x] /. First[sol1]
```

which yields the output

$$\text{InterpolatingFunction}[\{0., 2.\}, <>][x]$$

Now the approximate solution **ya1[x]** can be evaluated for any particular $x \in [0, 2]$, and hence can be graphed on the interval $0 \leq x \leq 2$. For example, typing **ya1[1.5]** yields the output 1.80277. Note that the exact value is $y(1.5) = \sqrt{13}/2 = 1.802775637\ldots$, so that **ya1[1.5]** is an approximate value with six correct digits.

The command

```
TableForm[Table[{x, ya1[x], N[Sqrt[x^2 + 1], 10]},
  {x, 0, 2, 0.2}]]
```

produces a table of numerical solution values and exact solution values at the points $0, 0.2, \ldots, 1.8, 2$. The result is shown in Table 1.

x	ya1[x]	$y(x)$
0	1.	1.
0.2	1.0198	1.019803903
0.4	1.07703	1.077032961
0.6	1.16619	1.166190379
0.8	1.28062	1.280624847
1.0	1.41422	1.414213562
1.2	1.56205	1.562049935
1.4	1.72046	1.720465053
1.6	1.88679	1.886796226
1.8	2.05912	2.059126028
2.0	2.23607	2.236067977

Table 1

We see that we have at least five correct digits in all cases. **NDSolve** attempts to produce an approximate solution with error less than or equal to 10^{-6}. See *Controlling the Error in NDSolve* and *Reliability of Numerical Methods* later in this chapter.

A graph of the approximate solution can be obtained with the command

```
Plot[ya1[x], {x, 0, 2}]
```

The result is shown in Figure 1.

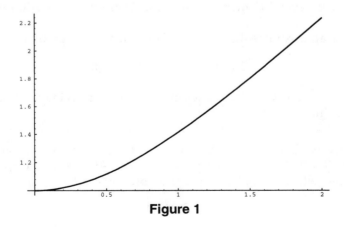

Figure 1

EXERCISE. Display both **ya1[x]** and the exact solution $y(x) = \sqrt{x^2 + 1}$ on the same graph. Can you distinguish the two curves?

When using **NDSolve**, the interval on which we calculate an approximate solution does not have to be of the form $[x_0, b]$; in particular, a can be less than x_0.

We have illustrated **NDSolve** on the initial value problem (1), which can easily be solved explicitly. We can just as readily apply **NDSolve** to any first order initial value problem. Consider, for example,

$$\frac{dy}{dx} = x + y^2, \qquad y(0) = 1, \tag{2}$$

which cannot be solved in terms of elementary functions. The commands

```
ivp2 = {y'[x] == x + y[x]^2, y[0] == 1}
sol2 = NDSolve[ivp2, y[x], {x, 0, 1}]
ya2[x_] = y[x] /. First[sol2]
```

will produce an approximate solution.

EXERCISE. The solution $y(x)$ to (2) is increasing for x positive, and there is a finite value x^* such that $\lim_{x \to x^*} y(x) = \infty$; the interval of existence of the solution is $(-\infty, x^*)$. Execute the previous two commands and then graph **ya2[x]**. Does **NDSolve** reveal the value of x^*?

REMARK. Although we have focused on first order equations in this chapter and in Chapter 5, both **NDSolve** and **DSolve** work on higher order equations and systems of equations. See Chapter 8, Chapter 12, and the sample solutions for examples.

Some Numerical Methods

In order to give some idea about how **NDSolve** calculates approximate values, we now discuss several numerical methods. These methods will be applied to problem (1).

We want to approximate the solution of the initial value problem

$$\frac{dy}{dx} = f(x, y), \qquad y(x_0) = y_0$$

on an interval $x_0 \le x \le b$. For n a positive integer, we divide the interval into n parts using points

$$x_0 < x_1 < x_2 < \cdots < x_n = b.$$

For simplicity, we assume that each part has the same width, or *step size*, $h = x_{i+1} - x_i = (b - x_0)/n$; therefore $x_i = x_0 + ih$. At each point x_i we seek an approximation, which we call y_i, to the true solution $y(x_i)$ at x_i:

$$y_i \approx y(x_i).$$

The Euler Method

The simplest numerical solution method is due to Euler and is based on the tangent line approximation to a function. Given the initial value y_0, we define y_i recursively by

$$y_{i+1} = y_i + h\,f(x_i, y_i), \qquad i = 0, 1, \ldots, n-1.$$

This formula is derived as follows:

$$
\begin{aligned}
y(x_{i+1}) &\approx y(x_i) + hy'(x_i), && \text{by the tangent line approximation} \\
&= y(x_i) + hf(x_i, y(x_i)), && \text{using the differential equation} \\
&\approx y_i + hf(x_i, y_i), && \text{since } y_i \approx y(x_i) \\
&= y_{i+1}.
\end{aligned}
$$

The approximations $y(x_i) \approx y_i$ should become better and better as h is taken smaller and smaller.

EXAMPLE. Consider the initial value problem (1):

$$\frac{dy}{dx} = \frac{x}{y}, \qquad y(0) = 1.$$

We wish to approximate $y(0.3)$ using the Euler Method with step size $h = 0.1$ and three steps. We find

$$
\begin{aligned}
&x_0 = 0, \quad x_1 = 0.1, \quad x_2 = 0.2, \quad x_3 = 0.3 \\
&y_0 = 1 \\
&y_1 = y_0 + hf(x_0, y_0) = y_0 + hx_0/y_0 = 1 \\
&y_2 = y_1 + hx_1/y_1 = 1.01 \\
&y_3 = 1.0298.
\end{aligned}
$$

Thus

$$y(0.3) = \sqrt{1.09} = 1.044030\ldots \approx y_3 = 1.0298$$

$$\text{Error} = |y(0.3) - y_3| = 0.0142\ldots.$$

Next use $h = 0.05$ and 6 steps:

$$
\begin{aligned}
y_0 &= 1 \\
y_1 &= 1 \\
y_2 &= 1.0025 \\
y_3 &= 1.0075 \\
y_4 &= 1.0149 \\
y_5 &= 1.0248 \\
y_6 &= 1.0370
\end{aligned}
$$

$$\text{Error} = |y(0.3) - y_6| = 0.0070\ldots.$$

If the initial value problem is "sufficiently smooth", then for the Euler Method one can show that the error in stepping from x_0 to any x_j in the interval $x_0 \le x \le b$ satisfies

$$\text{Error} \le Ch,$$

where C is a constant that depends on $f(x, y)$, its partial derivatives, the initial condition, and the interval, but not on h. Moreover, it can be shown that the error is actually proportional to h. Because of this, the Euler Method is called a *first order method*. Note that in the example, cutting the step size in half had the effect of cutting the error approximately in half, as expected for a first order method.

It is also useful to know the *local error*, which is the error made in one step. The error discussed in the previous paragraph is called *global error*, to distinguish it from local error. Suppose we are stepping from x_j to x_{j+1} and we denote the local error by e_{j+1}. To estimate the size of e_{j+1} we assume that $y_j = y(x_j)$, *i.e.*, that the approximation is exact at x_j. Using Taylor's formula it is easily shown that, for the Euler Method,

$$e_{j+1} = y(x_{j+1}) - y_{j+1} = \frac{1}{2}y''(\overline{x}_j)h^2,$$

where $x_j < \overline{x}_j < x_{j+1}$. Thus the local error is proportional to h^2. Because local error provides a simple comparison of methods and is used in the design of numerical solution software, we will state the local error for each method we discuss.

The relation between local error and global error can be intuitively understood in the following way. In stepping from x_0 to x_j, we make j local errors, each of which is proportional to h^2. Since $j \le n$ we thus have an accumulated error that is no greater than a constant times

$$nh^2 = \frac{b - x_0}{h}h^2 = (b - x_0)h;$$

i.e., the global error is bounded by a constant times h.

The Euler Method can be implemented using *Mathematica* as follows:

```
EulerMethod[f_, {x0_, y0_}, h_, n_] :=
    (EMStep[{x_, y_}] := N[{x + h, y + h*f[x, y]}];
        NestList[EMStep, {x0, y0}, n])
```

To apply this program to the initial value problem (1), we first define the function $f(x, y) = x/y$ and then run the routine. So,

```
f[x_, y_] := x/y
em = EulerMethod[f, {0, 1}, 0.1, 10]
```

will produce a list of approximate solution values at $x = 0, 0.1, 0.2, \ldots, 1$. These can be displayed using the command **TableForm[em]**. Note that **EulerMethod** begins by defining a new function, called **EMStep**, that performs a single Euler Method step. Then it uses **NestList** to iterate **EMStep**.

To approximate the solution at a point x that is not an x_i, we would have to *interpolate*. A simple way to do this is to "connect the dots" by drawing straight line segments from each point (x_i, y_i) to the next point (x_{i+1}, y_{i+1}); thus $y_a(x)$ would be the piecewise linear function connecting the computed points (x_i, y_i). To graph $y_a(x)$, type **ListPlot[em, PlotJoined -> True]**.

The Runge-Kutta Method

Given the initial value y_0, the fourth order Runge-Kutta Method is defined recursively by

$$y_{i+1} = y_i + \frac{h}{6}(k_1 + 2k_2 + 2k_3 + k_4),$$

where

$$k_1 = f(x_i, y_i)$$
$$k_2 = f(x_i + \frac{h}{2}, y_i + \frac{h}{2}k_1)$$
$$k_3 = f(x_i + \frac{h}{2}, y_i + \frac{h}{2}k_2)$$
$$k_4 = f(x_{i+1}, y_i + hk_3).$$

Note that a weighted average of the "slopes" k_1, k_2, k_3, k_4 is used in the formula for y_{i+1}.

EXAMPLE. Consider problem (1) and again find an approximation to $y(0.3)$, but now let $h = 0.3$ and take just one step. Then

$$k_1 = f(0, 1) = 0$$
$$k_2 = f(0 + 0.15, 1 + 0.15(0)) = 0.15$$
$$k_3 = f(0.15, 1 + 0.15(0.15)) = 0.146699$$
$$k_4 = f(0.3, 1 + 0.3(0.146699)) = 0.287354$$
$$y_1 = 1 + \frac{0.3}{6}(0 + 2(0.15 + 0.146699) + 0.287354)$$
$$= 1.044038 \approx y(0.3).$$

So Error $= |y(0.3) - y_1| = 0.000007$.... We see that the error in the Runge-Kutta Method with $h = 0.3$ is much less than the error in the Euler Method with $h = 0.05$.

The local error for the Runge-Kutta Method is proportional to h^5.

Adams-Moulton Methods

We discuss two methods from the Adams-Moulton family. The second order Adams-Moulton Method is defined by the formula

$$y_{i+1} = y_i + \frac{h}{2}(f(x_i, y_i) + f(x_{i+1}, y_{i+1})).$$

We start with an approximation y_i to $y(x_i)$, and use this formula to determine an approximation y_{i+1} to $y(x_{i+1})$. Since y_{i+1} appears on both sides of the equation, we will have to solve the equation for y_{i+1}. Because of this feature, the method is called *implicit*. The Euler and Runge-Kutta methods, in contrast, are *explicit*. The local error for the second order Adams-Moulton Method is proportional to h^3.

The fourth order Adams-Moulton Method is

$$y_{i+1} = y_i + \frac{h}{24}(f(x_{i-2}, y_{i-2}) - 5f(x_{i-1}, y_{i-1}) + 19f(x_i, y_i) + 9f(x_{i+1}, y_{i+1})).$$

This method is also implicit. One needs to know y_{i-2}, y_{i-1}, and y_i in order to use this formula to determine y_{i+1}. Because of this feature, the method is called a *multistep method*, specifically a *three-step method*. The previous methods are *one-step methods*. To use a multistep method, one must calculate a few initial values using some other (one-step) method. The local error for the fourth order Adams-Moulton Method is proportional to h^5.

There are Adams-Moulton Methods of any order. Adams-Moulton Methods are sometimes called Implicit Adams Methods.

EXAMPLE. Consider problem (1). Given the exact values for y_0, y_1, y_2, we use the Adams-Moulton Method with $h = 0.1$ to determine y_3. We can use the exact values since we have a formula for the solution, and we do so to simplify the example. As pointed out above, however, one would typically have to calculate these values using some other method.

For y_0, y_1, and y_2 we have

$$y_0 = 1, y_i = \sqrt{1.01} = 1.00499, \text{ and } y_2 = \sqrt{1.04} = 1.01980.$$

Then y_3 is determined from

$$y_3 = y_2 + \frac{h}{24}(x_0/y_0 - 5x_1/y_1 + 19x_2/y_2 + 9x_3/y_3)$$

$$= 1.01980 + \frac{0.1}{24}\left(0/1 - 5(0.1/1.00499) + 19(0.2/1.01980) + 9(0.3/y_3)\right)$$

$$= 1.03326 + 0.01125/y_3.$$

This simplifies to the quadratic equation

$$y_3^2 - 1.03326y_3 - 0.01125 = 0,$$

whose solutions are 1.04403 and -0.0107755. The positive solution is the desired solution, so $y_3 = 1.04403$. Since

$$y(0.3) = \sqrt{1.09} = 1.0440306\ldots \approx 1.04403,$$

our numerical answer has 6 correct digits.

Backwards Differentiation Formulas

The fourth order Backwards Differentiation Formula is

$$y_{i+1} = \frac{1}{25}(-3y_{i-3} + 16y_{i-2} - 36y_{i-1} + 48y_i + 12hf(x_{i+1}, y_{i+1})).$$

This is an implicit, four-step method. Its local error is proportional to h^5. There are Backwards Differentiation Formulas of any order. Backwards Differentiation Formulas are sometimes referred to as the Gear Formulas.

The methods we have discussed can be applied to any first order equation. Furthermore, they can be applied using different *step sizes* and different *orders* at each step.

Inside NDSolve

Software for the numerical solution of ordinary differential equations can be based on the methods we have presented, as well as on other methods. By default, **NDSolve** uses the Adams-Moulton Methods and the Backwards Differentiation Formulas. It varies the step size and the order of the methods or formulas, choosing the step size and the order at each step to attempt to ensure that the desired accuracy is attained. All efficient modern numerical ODE solvers use variable step size methods, and many use variable order methods.

In addition to choosing step sizes and orders, **NDSolve** chooses between the Adams-Moulton Methods and the Backwards Differentiation Formulas, depending on the problem. Initial value problems can be classified as *stiff* or *non-stiff*. **NDSolve** attempts to test for stiffness; it then uses the Adams-Moulton Methods for non-stiff problems and the Backwards Differentiation Formulas for stiff problems. Stiff problems are generally more difficult to solve numerically, and the Backwards Differentiation Formulas are particularly effective for such problems. For additional information on numerical methods and software we refer to L. Shampine, **Numerical Solution of Ordinary Differential Equations**, Chapman & Hall, 1994 and D. Kahaner, C. Moler, and S. Nash, **Numerical Methods and Software**, Prentice Hall, Inc., 1989.

NDSolve produces approximate solution values at a sequence of points. Then it interpolates between the points it has computed, not by straight lines, but by polynomials, thereby producing a smooth approximate solution $y_a(x)$. **NDSolve** attempts to perform both approximation processes—the calculation of the approximate solution values and the interpolation process—in such a way that the total error in both processes, $|y(x) - y_a(x)|$, does not exceed the desired level, by default 10^{-6}.

The type of error discussed above is called *discretization error*. In addition, there is *round-off error*, which arises because the computer uses a fixed, finite number of digits. Letting $\tilde{y}_a(x)$ denote the actual computed approximate solution, which includes round-off error, the total error can be written

$$y(x) - \tilde{y}_a(x) = (y(x) - y_a(x)) + (y_a(x) - \tilde{y}_a(x))$$
$$= \text{Discretization Error} + \text{Round-off Error}.$$

Since most computers that run *Mathematica* carry 16 digits, the major portion of the error will be the discretization error. Thus, for most problems, round-off error can be safely ignored. For this reason we will not distinguish between $y_a(x)$ and $\tilde{y}_a(x)$.

You can use the **Method** option with **NDSolve** to select other methods. The setting **Method** $->$ **Adams** specifies a method based on the Adams-Moulton Methods. **Method** $->$ **Gear** specifies a method based on the Gear, or Backwards Differentiation, Formulas. **Method** $->$ **RungeKutta** leads to the Runge-Kutta-Fehlberg Method, which is based on fourth and fifth order Runge-Kutta formulas. The default setting, **Method** $->$ **Automatic**, specifies the combination of the Adams-Moulton Methods and the Backwards Differentiation Formulas discussed above.

REMARK. With numerical methods as with qualitative methods, we obtain information on the solution—quantitative in the one case and qualitative in the other—without the use of a solution formula.

Controlling the Error in NDSolve

As indicated above, **NDSolve** attempts to provide an approximate solution with error approximately $\leq 10^{-6}$. A more accurate approximate solution can be obtained by using certain options to **NDSolve**. For example,

```
sol3 = NDSolve[{y'[x] == x/y[x], y[0] == 1}, y[x],
 {x, 0, 2}, AccuracyGoal -> 15, PrecisionGoal -> 15,
 WorkingPrecision -> 25]
ya3[x_] = y[x] /. First[sol3]
N[ya3[1.5], 16]
```

produces 1.802775637731983, which is a 14-digit approximation to the exact value $\sqrt{13}/2$.

AccuracyGoal specifies absolute error, and **PrecisionGoal** specifies relative error. With the settings **AccuracyGoal** $->$ **a** and **PrecisionGoal** $->$ **p**, where **a** and **p** are positive integers, **NDSolve** attempts to calculate $y(x)$ with error less than $10^{-a} + |y(x)|10^{-p}$.

The setting **WorkingPrecision** $->$ **n** determines the number of digits that **NDSolve** uses in its calculations. Without these options, *Mathematica* uses the default settings:

WorkingPrecision = Normal machine precision for the computer
AccuracyGoal = **PrecisionGoal** = **WorkingPrecision** - 10

So for any computer that has a machine precision of 16 digits, the default options lead to approximately 6-digit accuracy. This will be satisfactory for most of the calculations in this course; but see *Reliability of Numerical Methods* below.

When you use *Mathematica* to calculate an approximate numerical solution to an initial value problem on an interval $[a, b]$, you should be aware that the solution may not exist over the entire interval. Naturally, *Mathematica* can only calculate the solution over the largest subinterval of $[a, b]$ on which it exists. See the second exercise in this chapter to learn how **NDSolve** reports the answer in this situation.

Now suppose the solution does exist over the entire interval $[a, b]$. In attempting to calculate the solution with the desired accuracy, **NDSolve** will use up to 1000 steps. On certain problems this limit will be reached before the solution has been calculated over the entire interval, and **NDSolve** will return a message indicating this. You can then change the default setting to a higher number, say 1100, with the option **MaxSteps** $->$ **1100**, and obtain the solution on a (possibly) larger interval.

EXERCISE. Attempt to calculate the solution to

$$\frac{dy}{dx} = y\sin(x^2), \qquad y(0) = 1$$

over the interval $[0, 19]$, first using the default option (**MaxSteps** $->$ **1000**), and then using **MaxSteps** $->$ **1100**. What difference do you observe? Find a value for **MaxSteps** that allows the solution to be calculated over the entire interval $[0, 19]$. Graph the solution.

Reliability of Numerical Methods

We have claimed that **NDSolve**, when used with the default options, leads to approximately 6-digit accuracy. More precisely, the default options ensure that the *local error*, *i.e.*, the error made in a step of length h, is approximately $10^{-6} \times h$. One is generally more interested in the *global error*, defined above to be the cumulative error committed in taking the necessary number of steps to get from the initial point x_0 to b. This error cannot be controlled completely by the numerical method because it depends on the differential equation as well as on the numerical method. The key issue is whether the differential equation is stable or unstable; these terms were defined in Chapter 5.

We have seen equations for which **NDSolve** gives accurate results. Now we look at an example where **NDSolve** gives poor results.

EXAMPLE. Consider the initial value problem

$$\frac{dy}{dx} - y = -3e^{-2x}, \qquad y(0) = 1.$$

The exact solution is $y(x) = e^{-2x}$; we plot it in Figure 2.

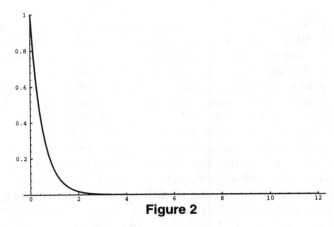

Figure 2

Next we find the numerical solution using **NDSolve**, and plot it together with the exact solution:

```
sol4 = NDSolve[{y'[x] - y[x] == -3Exp[-2x], y[0] == 1},
  y[x], {x, 0, 14}]
ya4[x_] = y[x] /. First[sol4]
Plot[{Exp[-2x], ya4[x]}, {x, 0, 12}, PlotRange -> {0, 1}]
```

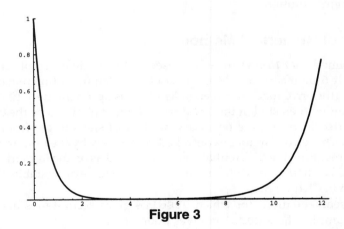

Figure 3

We see from Figure 3 that the graphs of **ya4** and y are indistinguishable from 0 to about 6, but then the curves separate sharply. Their values at $x = 12$, for instance, are very different (**ya4[12]** $= 0.76\ldots$ and $y(12) = e^{-24} = 3.77\ldots \times 10^{-11}$).

How can this failure be explained? The solution to our differential equation with initial condition $y(0) = 1 + \epsilon$ is $y(x) = e^{-2x} + \epsilon e^x$. Figure 4 shows a plot of the solutions corresponding to initial values $1 - (0.5 \times 10^{-5}), 1 - (0.25 \times 10^{-5}), 1, 1 +$ $(0.25 \times 10^{-5}), 1 + (0.5 \times 10^{-5})$.

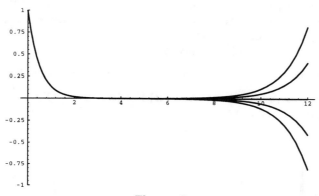

Figure 4

We see that as x increases, the solutions with initial values slightly different from 1 separate very sharply from the solution with initial value exactly 1; those with initial values greater than 1 approach $+\infty$, and those with initial values less than 1 approach $-\infty$. Now **NDSolve**, like any numerical method, introduces errors. These errors—whether discretization errors or round-off errors—have caused the numerical solution to jump to a solution that started just above 1. Once on such a solution, the numerical solution will follow it or another such solution. Since these solutions approach $+\infty$, the numerical solution does likewise.

Could we have anticipated this failure? Recall the discussion of stable and unstable differential equations in Chapter 5. The solutions of unstable equations are very sensitive to their initial values, and hence a numerical method will have trouble following the solution it is attempting to calculate. Recall also that an equation $dy/dx = f(x, y)$ is unstable if $\partial f/\partial y > 0$, and stable if $\partial f/\partial y \leq 0$. For our example, $f(x, y) = y - 3e^{-2x}$ and $\partial f/\partial y = 1 > 0$. So we are trying to approximate the solution of an unstable differential equation, and we should not be surprised that we have trouble, especially over long intervals. (By contrast, note that for the initial value problem (1) above, $\partial f/\partial y = -x/y^2 < 0$ for $x > 0$.) All this suggests caution when dealing with unstable equations. In fact, caution is always recommended when modeling physical problems with unstable equations, especially over long intervals (*cf.* the discussion in Chapter 5).

In assessing the reliability of a numerically computed solution there is another test one can make. Suppose we have made a calculation using the default options, as we have done with our example. Then one can do another calculation, using a more stringent accuracy requirement, and compare the results. If they are nearly the same, one can have reasonable confidence that they are both accurate; but if they differ substantially, then one should suspect that the first calculation is not accurate.

Let's do this with our example:

```
sol5 = NDSolve[{y'[x] - y[x] == -3Exp[-2x], y[0] == 1},
  y[x], {x, 0, 14}, AccuracyGoal -> 8, PrecisionGoal -> 8,
  WorkingPrecision -> 18]
ya5[x_] = y[x] /. First[sol5]
```

Next we plot both numerical solutions together with the exact solution:

```
Plot[{Exp[-2x], ya4[x], ya5[x]}, {x, 0, 14},
  PlotRange -> {0, 1}]
```

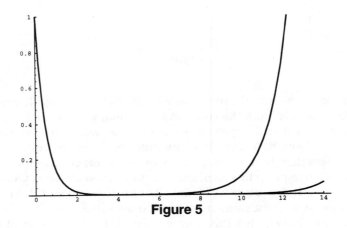

Figure 5

We see from Figure 5 that the second numerical solution is accurate over a longer interval, out to about 10, whereas the first was accurate out to about 6. In particular, the two numerical solutions differ after about 6. Even if we didn't know the exact solution, we could have concluded that the first numerical solution is accurate out to, but not beyond, about 6.

In summary, we have learned that local accuracy—the accuracy that a numerical method can control—leads to reasonable global accuracy if the differential equation is stable, and that the reliability of a numerical solution can also be checked by computing another numerical solution with a more stringent accuracy requirement.

Chapter 8

Features of *Mathematica*

This chapter is a continuation of Chapter 3, *Doing Mathematics with Mathematica*. We describe some of the more advanced features of *Mathematica*, focusing on the *Mathematica* commands and techniques that are most useful for studying differential equations.

Functions

Mathematica distinguishes between a *function* of a variable and an *expression* involving the variable. In order to use *Mathematica* effectively, you must learn to make the same distinction. You should review the section *User-defined Functions and Expressions* in Chapter 3 to refresh your memory on the difference between functions and expressions.

The simplest way to define a function is via the assignment method. For example, to define $f(x)$ to be the cubic polynomial $x^3 - 2x^2 + 1$, type

```
In[1]:=  f[x_] = x^3 - 2x^2 + 1
```
$$\text{Out[1]= } 1 - 2x^2 + x^3$$

The argument of the function must be enclosed in square brackets, and on the left-hand side of the equal sign the argument must be followed by an underscore.

A function can be evaluated with either a numeric or a symbolic argument:

```
In[2]:=  f[-1]
```
$$\text{Out[2]= } -2$$

```
In[3]:=  f[u^2]
```
$$\text{Out[3]= } 1 - 2u^4 + u^6$$

The result of this evaluation is the expression $1 - 2u^4 + u^6$. This is a subtle point: the symbol **f** is the name of a *function*; the symbol **f[u]** is the name of an *expression* involving **u**.

An alternative way of defining a function is to delay evaluation by using the "colon-equal" sign. For example,

```
In[4]:=  f[x_] := x^3 - 2x^2 + 1
```

The difference between the two methods is the following. In the first (without the colon), the right-hand side of the definition is evaluated at the time the definition is made. In the second (with the colon), the right-hand side of the definition is evaluated only when the value of the function is requested. This is illustrated by *Mathematica*'s lack of response to the second definition, which indicates that *Mathematica* is holding the definition in reserve for evaluation at a later time. Here is an example to illustrate this distinction.

```
In[5]:=  expr = x^2 + 3
```
$$\text{Out}[5] = \ 3 + x^2$$

```
In[6]:=  g[x_] = expr
```
$$\text{Out}[6] = \ 3 + x^2$$

```
In[7]:=  h[x_] := expr
```

```
In[8]:=  g[2]
```
$$\text{Out}[8] = \ 7$$

```
In[9]:=  h[2]
```
$$\text{Out}[9] = \ 3 + x^2$$

What happens here is that when *Mathematica* evaluates a function defined with the := notation, it *first* substitutes the argument (in this case x) into the expression on the right-hand side, and *then* evaluates the right-hand side. Since x does not appear explicitly as a variable on the right-hand side in the definition of h, the function we define this way is actually independent of x. Thus h has been defined to be a function that returns as its *constant* value the expression *expr*. The point of this discussion is that the distinction between = and := can be important depending upon when you actually want evaluation to occur. A crucial example appears later when we discuss **NDSolve**.

Finally, we can define functions of several variables. For example, we could define **phi[x_,y_] = x^2 + y^2**.

Pure Functions

In the examples above, each of the functions we defined was given a name—*f*, *g*, *h*, or *phi*. *Mathematica* provides a way to define functions that are unnamed, or *pure*. There are two types of syntax for pure functions. Here are the two ways to define the pure function that takes x to $x^3 - 2x^2 + 1$.

```
Function[x, x^3 - 2x^2 + 1]
```

```
#^3 - 2 #^2 + 1 &
```

In the latter syntax, the number sign (#) is a stand-in for the variable, and the ampersand (&) indicates that the expression is a pure function. The only time we will use pure functions in this book is as scale functions for the **PlotVectorField** command, but *Mathematica* occasionally reports its answers in pure function form. For example, if **f** has been defined as in the preceding section, then *Mathematica* reports the derivative in pure function format if you type

In[10]:= **f'**

Out[10]= $-4 \, \#1 + 3 \, \#1^2 \&$

The "1" after the number sign is there to indicate the first variable. There is only one variable here, but the syntax allows for functions of several variables. You can, however, avoid the pure function format in this case by typing

In[11]:= **f'[x]**

Out[11]= $-4x + 3x^2$

Suppressing Output

Certain commands produce output that might be considered superfluous. For example, when you assign a value to a variable, *Mathematica* will echo the value. The output of a *Mathematica* command can be suppressed by putting a *semicolon* after the command. Semicolons can also be used to separate a string of *Mathematica* commands when you are only interested in the output of the final command. Ending a statement with a semicolon has no effect on what *Mathematica* does internally; it only affects what you see in the Notebook. A semicolon after a graphics command will not suppress the graphic, but it will suppress the superfluous *Output* cell that ordinarily follows a graphic.

Clearing Values

One of the most common sources of errors in using *Mathematica* is the failure to keep track of variable or function definitions. For example, suppose that during the course of a *Mathematica* session you type

In[12]:= **k = 1;**

You keep working, and twenty minutes later you try to define a function of two variables by typing

In[13]:= **g[x_, k_] = x + k;**
         g[3, 4]

Out[13]= 4

The answer should have been $3 + 4 = 7$. But the function **g[x, k]** is not really **x + k**; rather it's **x + 1** since **k** has previously been set to 1. In this simple case you might have noticed that **g** wasn't what it should have been, but in a complicated expression this type of error is difficult to detect.

The simplest way to avoid this problem is to clear a variable either before using it or immediately after using it. For example, you can avoid errors like the one in the preceding example by typing

In[14]:= **Clear[g, x, k]**
         g[x_, k_] = x + k

Out[15]= x + k

Another way to avoid the error is by using **:=** instead of **=**.

In[16]:= **k = 1;**
         g[x_, k_] := x + k
         g[3, 4]

Out[17]= 7

In this example, because we have used **:=** to define the function, *Mathematica* does not evaluate the right-hand side until the function is actually invoked, at which time it uses the value of the variable specified in the function call rather than the previously defined value. In the earlier case, *Mathematica* evaluates the right-hand expression immediately and substitutes, once and for all, the value 1 for **k**. For this reason, it is often safer to use **:=** rather than **=** when defining functions.

If you typed something like **q'[t] = t^2** (which is easy to do accidentally when using **DSolve**), then **Clear[q']** doesn't work, but **Clear[q]** does.

You can always find out the current value of a function or variable by typing a question mark before the function or variable name. For example, **?f** will show you the current value or definition of **f**.

You should be aware that if you have several Notebooks open in a single *Mathematica* session, the definitions you make in one Notebook will carry over to the other Notebooks. This happens because the underlying *Mathematica* process, or *kernel*, is the same for all the Notebooks. This also accounts for the fact that In/Out numbering carries over from Notebook to Notebook in a single *Mathematica* session, so if you enter input in one Notebook and the input is labeled *In[n]*, then the next input will be labeled *In[n+1]*, even if it is in a different Notebook.

Managing Memory

We have noted that *Mathematica* stores in computer memory *every* input statement and *every* output (that's what it's doing when it assigns those labels *In[n]* and *Out[n]*). In a long *Mathematica* session, this uses a lot of computer memory and might lead to a crash if the available memory is exhausted. The simplest way to ameliorate this problem is to reset the line numbering to zero. To do this, simply type **$Line=0**. The In/Out numbering will be reset, and subsequent inputs and outputs will be written over the old ones instead of using new memory space.

Another way to free up memory is to quit and restart the kernel. We described how to do this in Chapter 4. You should probably quit and restart the kernel each time you start a new problem.

The Replacement Operator and Transformation Rules

We have seen that a function can be evaluated at various values of the independent variable. A similar effect can be accomplished with expressions by using the *replacement operator* **/.**, together with a *transformation rule*. For example,

```
In[18]:= (x^2 - 1) /. x -> 5
```

```
Out[18]= 24
```

This input means: in the expression $x^2 - 1$, replace x by 5. The notation **x -> 5** is an example of a transformation rule.

The replacement operator is extremely useful and versatile. What it does is to substitute for one or more variables in an expression using one or more transformation rules. Here is another example:

```
In[19]:= ((x^2 - 3*y)/z) /. {x -> a, y -> b, z -> c}
```

$$Out[19]= \frac{a^2 - 3b}{c}$$

A general feature of *Mathematica*'s various "solve" routines is that the solutions are reported in the form of transformation rules. For example, the simple quadratic equation $x^2 - 1 = 0$ has the two solutions $x = 1$ and $x = -1$.

```
In[20]:= sol = Solve[x^2 - 1 == 0, x]
```

$$Out[20]= \{\{x \to 1\}, \{x \to -1\}\}$$

Mathematica's response is a list consisting of two transformation rules. (We have named this list **sol** so we can use it later.) Note that *Mathematica* does not report these solutions in the form "$x = 1$" and "$x = -1$", because those two assignments would be in conflict. In fact, *Mathematica* *never* assigns values to variables unless specifically directed to do so. Transformation rules are *Mathematica's* way of steering clear of assigning values to variables.

Now let us use the transformation rules that we generated with the **Solve** command. To do this, we have to extract the transformation rules from the list. If we want to use the rule **x** $\rightarrow$ **1** on the expression $x^2 + x$, we type

> In[21]:= **x^2 + x /. First[sol]**

> Out[21]= 2

To use the rule **x** $\rightarrow$ **-1**, we type

> In[22]:= **x^2 + x /. Last[sol]**

> Out[22]= 0

The most common use of the replacement operator is to transform expressions that have been previously typed by the user or generated by *Mathematica*. Here is a more complicated example, in which we find the function that solves the initial value problem $y' = xy$, $y(0) = 1$.

> In[23]:= **sol1 = DSolve[{y'[x] == x*y[x], y[0] == 1}, y[x], x]**

> Out[23]= $\{\{y[x] \rightarrow E^{\frac{x^2}{2}}\}\}$

> In[24]:= **Clear[f]**
>   **f[x_] = y[x] /. First[sol1]**

> Out[25]= $E^{\frac{x^2}{2}}$

DSolve reports its result as a transformation rule. We have used this transformation rule to define a function f, which is the solution to the initial value problem.

Equations vs. Assignments

In *Mathematica*, an *equation* is different from an *assignment*. For example, **x == 2** is an equation, whereas **x = 2** or **x := 2** is an assignment. Compare the effects of entering an equation and an assignment.

> In[26]:= **x == 2**

> Out[26]= $x == 2$

> In[27]:= **x = 2**

> Out[27]= 2

Note that an equation has no effect on the value of the variables in the equation. Names can even be assigned to equations.

```
In[28]:= niceeqn =   energy == m*c^2
```

Out[28]= energy $== c^2 m$

Differentiation

The distinction between functions and expressions becomes important when differentiating. *Mathematica* has two commands for differentiating. The **D** command operates on expressions.

```
In[29]:= Clear[g, x]
         g[x_] := x^2
         D[g[x], x]
```

Out[31]= $2x$

Notice that the output of **D** is another expression. The syntax for second derivatives is **D[g[x], {x, 2}]**, and for *n*th derivatives, **D[g[x], {x, n}]**. When differentiating an expression with **D**, you must state explicitly which variable to use.

As we observed in the section on pure functions, the *prime* operator differentiates a function and returns a pure function. One place where you must use the *prime* format is to specify initial values for derivatives in **DSolve** or **NDSolve**. So, to solve the second order initial value problem $y'' = 9y + 2x$, $y(0) = 1$, $y'(0) = 0$, you could type

```
In[32]:= diffeqn = D[y[x], {x, 2}] == 9*y[x] + 2*x
```

Out[32]= $y''[x] == 2x + 9y[x]$

```
In[33]:= initialpos = y[0] == 1;
         initialvel = y'[0] == 0;
```

```
In[34]:= DSolve[{diffeqn, initialpos, initialvel}, y[x], x]
```

Out[34]= $\left\{\left\{y[x] \rightarrow \dfrac{\frac{25}{6} + \frac{29 e^{6x}}{6} - 2 e^{3x} x}{9 e^{3x}}\right\}\right\}$

Graphics

We will encounter the following six plotting commands in this book: **Plot**, **ListPlot**, **ContourPlot**, **PlotVectorField**, **Show**, and **ParametricPlot**. We briefly describe each of them in this section.

The most important thing to keep in mind for the various plotting commands is the type of *Mathematica* objects that the command is designed to plot. For example, the **Plot** command plots expressions involving a single variable. On the other hand, **PlotVectorField** plots a list of two expressions involving two variables. If *Mathematica* gives you an error message or an implausible plot when you enter a plotting command, the first things to check are whether you are plotting the right kind of object for that particular command, and whether you are using precisely the right syntax for that object and that command.

Plot is the simplest and most versatile plotting command; it plots an expression or list of expressions involving a single variable over a specified interval. The vertical range of the plot can also be specified using the option **PlotRange**. For example, to plot the two functions $e^x \cos x$ and $x^3 - 3x + 2$ over the interval $[-3, 3]$ with range $[-10, 10]$, we type

```
In[35]:= plot1 = Plot[{Exp[x]*Cos[x], x^3 - 3x + 2},
         {x, -3, 3}, PlotRange -> {-10, 10}];
```

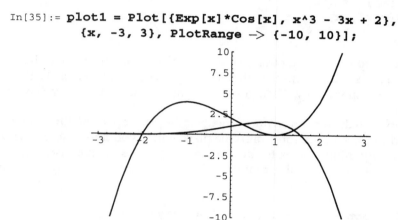

We have named this plot for future reference.

When *Mathematica* responds to a **Plot** command (or any of the plotting commands described below), it samples points according to a default setting. If curves are exceptionally oscillatory, or there are many curves to be generated by the command, the default setting may be inadequate. You can adjust it with the option **PlotPoints** -> **n**. The default setting for **Plot** is $n = 25$. Higher settings may improve your graphs, at the cost of an increase in computation time.

ListPlot takes a list of points (given as pairs of (x, y) coordinates) and plots them. The list may be given explicitly or it may be generated by the **Table** command. If you use the option **PlotJoined** -> **True**, then **ListPlot** will connect the plotted points with line segments.

```
In[36]:= plot2 = ListPlot[
         {{-3, 1}, {-1, 2}, {0.5, 8}, {2, 3}, {3, 2}},
         PlotJoined -> True];
```

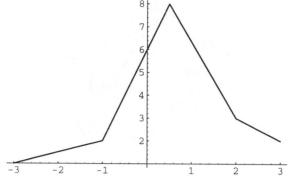

ContourPlot plots level curves of an expression involving two variables, *i.e.*, sets of points on which the expression is constant.

```
In[37]:= ContourPlot[x^2/2 + y^2, {x, -5, 5}, {y, -5, 5}];
```

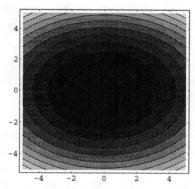

The shading in the plot represents different values of the function: the larger the value of the function, the lighter the color. Shading can be turned off using the option **ContourShading -> False**. The option **Contours -> n** produces *n* contours. The option **Contours -> {a, b, ...}** produces contours at levels $a, b, \ldots$.

Note that if you want to apply **ContourPlot** to a previously defined function **f**, you must explicitly enter the independent variables, as in **f[x, y]**, rather than just **f**. The same holds for all the plotting commands. (Another way of saying this is that the plotting routines plot *expressions*, and not *functions*.)

ParametricPlot is used to plot the trajectory traced out by a pair of expressions **{f[t], g[t]}** as the parameter **t** varies. It is useful for graphing solutions of systems of differential equations.

```
In[38]:= ParametricPlot[
         {Exp[-t/100]*Cos[t], Exp[-t/100]*Sin[t]},
         {t, 0, 100}, AspectRatio -> Automatic];
```

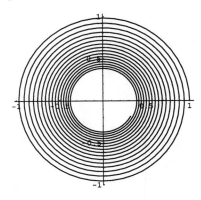

We have included the option **AspectRatio** $->$ **Automatic** to control the height-to-width ratio in the plot. The default ratio is the reciprocal of the so-called *Golden Ratio*, approximately 0.618, which is considered to be a ratio that is "pleasing" to the eye. The **Automatic** option instructs *Mathematica* to use the same scale on both axes.

The **PlotRange** option can be used to adjust both the horizontal and vertical range of the plot. The syntax is **PlotRange** $->$ **{{a, b}, {c, d}}**. This will restrict the plot to the rectangle $a \leq x \leq b$, $c \leq y \leq d$.

PlotVectorField is used for plotting vector or direction fields and was described in Chapter 6. It is in the **Graphics`PlotField`** package. The **ScaleFunction** option to the **PlotVectorField** command can be used to rescale the length of vectors in the plot. The default setting causes the vector field to be displayed with vectors having length proportional to the magnitude of the actual vectors. But often one is interested in the behavior of a vector field near critical points, where the vectors are very short. The default setting produces a plot that is relatively uninformative near critical points, because the vectors are so short that it is impossible to discern their direction. The simplest solution is to make all the vectors the same size. The option **ScaleFunction** $->$ **(1&)** does this. The ampersand is present because **ScaleFunction** expects a *pure function* as its argument; we have used the constant function 1. Note that the default setting of **AspectRatio** is **Automatic** for **PlotVectorField**; this can result in a thin graph if the x and y ranges of the plot have different lengths. In such cases you can use the option **AspectRatio** $->$ **1** to obtain a square graph.

The **Show** command is used to superimpose several graphs. We named two of the graphs above **plot1** and **plot2** so that we could use the **Show** command on them.

In[39]:= **Show[plot1, plot2];**

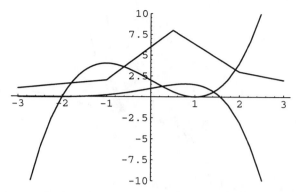

The **Show** command always puts the graphs on the same horizontal and vertical scales, so if you are going to use the **Show** command to superimpose two graphs, you may want to make sure that the plotting ranges of the two graphs are comparable.

Various other options are available for improving the quality of the pictures generated by the plotting commands. For example,

```
Plot[x^3 - 3, {x, 0, 2}, AxesLabel -> {"x", "y"}]
```

will cause the axes to be labeled. You can learn about many other options by typing **??Plot**. Some of the most useful are: **Axes -> False** (hides the axes), **Ticks -> None** (turns off the tick marks on the axes), **Frame -> True** (encloses the plot in a box), and **PlotStyle -> GrayLevel[0.5]** (causes the curve to be plotted in a shade of gray, with 0 representing black and 1 representing white, and numbers in between representing shades between black and white).

Finally, you can locate coordinates on your graphs much as you would on a graphics calculator. This is explained in the *Graphics* section of Chapter 4.

The Evaluate Command

Each of the plotting commands just described can be applied to lists of arguments rather than to a single argument. For example, in the preceding section we applied the **Plot** command to a list of two functions. It is also possible to pass a list of functions to a plotting command using the **Table** command, or some other *Mathematica* command. However, when you use a command inside of a plotting routine, you must use **Evaluate** to force evaluation of the command. For example, we could type

```
In[40]:= Plot[Evaluate[Table[x^c, {c, 1, 5}]], {x, -5, 5}];
```

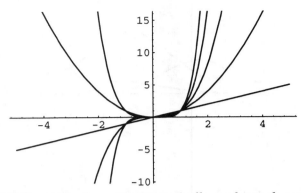

The reason *Mathematica* doesn't automatically evaluate the arguments to the plotting commands is this: In the command we typed above, there are two possible orders of evaluation. One could first create the list of five functions corresponding to the values $c = 1, \ldots, 5$, and then evaluate these functions at each value of x. Or, for each value of x, one could evaluate the function at (x, c), for each of the five values of c. This is a subtle difference, but in many cases the first order of evaluation is easier, and this is the order that *Mathematica* uses when you type the command above. Since the order of evaluation is ambiguous, *Mathematica* requires that you tell it explicitly what to do by using the **Evaluate** command.

Here is another example of the **Evaluate** command. Suppose we define

```
f[t_] := {{f1[t], g1[t]}, {f2[t], g2[t]}}
```

where the functions **f1[t], ..., g2[t]** have been previously defined. If we want to apply **ParametricPlot** to the list **f[t]**, then we must use **Evaluate** inside the plotting routine:

```
ParametricPlot[Evaluate[f[t]], {t, -3, 3},
 AspectRatio -> Automatic]
```

It is sometimes hard to tell whether **Evaluate** is necessary, but if you're having trouble getting your plots to work, you should try it. In our experience, it never hurts to include **Evaluate**, even when it is not necessary.

DSolve

We discussed **DSolve**, *Mathematica*'s symbolic ODE solver, in Chapter 5. Here we present some additional remarks about this command.

Here is an application of **DSolve** to find the general solution of a second order equation.

```
In[41]:= gensol = DSolve[y''[x] - y[x] == 0, y[x], x]
```
$$Out[41]= \left\{\left\{y[x] \rightarrow \frac{C(1)}{E^x} + E^x\, C(2)\right\}\right\}$$

You should recognize from the discussion earlier in this chapter that *Mathematica* has reported the solution as a *list* of transformation rules, even though there is only one element in the list. The output is a general solution, and the expressions **C[1]** and **C[2]** are arbitrary constants. These constants are "protected" variables, so we can't actually assign values to them. But we can replace them with particular values by using a transformation rule.

```
In[42]:= ygen[x_] = y[x] /. First[gensol];
         y0[x_] = ygen[x] /. {C[1] -> 1, C[2] -> -1}
```

Out[42]= $E^{-x} - E^x$

NDSolve

The function **DSolve** yields *symbolic* solutions of differential equations, *i.e.*, algebraic expressions that satisfy the original differential equation. **NDSolve**, on the other hand, generates approximate *numerical* solutions of differential equations. These solutions are really just lists of numbers corresponding to approximate values of the solution at particular values of the independent variable. The output of **NDSolve** can generally be treated as a function, however, because *Mathematica* interpolates between values in the list. In fact, the only essential difference you have to keep in mind when using **NDSolve** (as opposed to **DSolve**) is that **NDSolve** requires actual *numbers*, whereas **DSolve** can work with variables. In particular, you must specify a numerical initial condition when using **NDSolve**.

```
In[43]:= approxsol =
         NDSolve[{y'[x] == Sqrt[y[x] + x], y[1] == 3},
         y[x], {x, 1, 10}]
```

Out[43]= $\{\{y[x] \to \text{InterpolatingFunction}[\{1., 10.\}, <>][x]\}\}$

The output is different from that of **DSolve**—the transformation rule specifies an interpolating function rather than a specific algebraic expression. But we can manipulate the output in exactly the same way as we did for **DSolve**.

```
In[44]:= ya[x_] = y[x] /. First[approxsol]
```

Out[44]= $\text{InterpolatingFunction}[1., 10., <>][x]$

This defines a function **ya[x]**, which is an approximation to the actual solution of the initial value problem. Note that this function is defined on the interval [1, 10], which is the interval we specified when we executed the **NDSolve** command. We can plot it (with a command like **Plot[ya[x], {x, 1, 10}]**), or evaluate it at specific points (*e.g.*, **ya[3.45]** or **Table[ya[1 + i], {i, 0, 9}]**). *Mathematica* will extrapolate the function if you try to plot it or evaluate it outside the domain of definition.

It is possible to have the function depend on the initial data, but here the difference between **NDSolve** and **DSolve** becomes important. Since **NDSolve** works numerically rather than symbolically, it must have specific initial data to do its computations. Thus we must define the function as follows:

```
In[45]:= f[xval_, c_] := (y[x] /. First[NDSolve[
          {y'[x] == Sqrt[y[x] + x], y[1] == c}, y[x],
          {x, 1, 10}]]) /. x -> xval
```

Note that **f** is defined with **:=**. This is necessary to prevent *Mathematica* from evaluating **NDSolve** before specific initial data are supplied. When **f** is evaluated for particular values of $xval$ and c, the expression in parentheses computes the numerical solution of the differential equation for the initial condition $y(1) = c$; then the replacement operator and transformation rule **x -> xval** evaluates this solution at the particular value of $xval$.

Now, to plot several solution curves corresponding to different initial data, you can type

```
In[46]:= Plot[Evaluate[Table[f[x, c], {c, 1, 5}]],
          {x, 1, 10}];
```

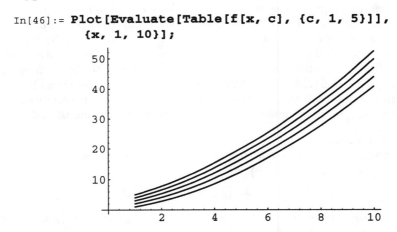

To compute values of **f** for specific choices of the initial data, *e.g.*, to evaluate the value of the solution function at $x = 2$ for initial data $c = 5$, you simply type **f[2, 5]**.

Troubleshooting

In this section, we list some advice for avoiding common mistakes. We also discuss warning messages and describe some techniques for recovering from errors.

The Most Common Mistakes

Here are some of the most common mistakes in using *Mathematica*.

(1) Failing to clear values. See *Clearing Values* above.

(2) Failing to keep track of In/Out numbering. This mistake occurs if you use **%** to refer to the output of the previous command. The previous command is the one most recently executed, not necessarily the preceding command in the Notebook. See *Referring to Previous Output* in Chapter 3.

(3) Using a command from a *Mathematica* package before the package has been loaded. See *Packages* in this chapter.

(4) Typing the wrong kind of "=" sign. The worst instance is typing "=" when you intend "==". This causes a value to be assigned to an expression, and this value must be explicitly cleared before you can proceed. Be particularly wary of this mistake when using **Solve**, **DSolve**, and **NDSolve**, all of which require an equation as one of the arguments.

(5) Mismatching braces or brackets. *Mathematica* usually catches this kind of mistake.

(6) Improperly using of built-in functions. For example, to use the natural logarithm function in *Mathematica*, you type **Log[x]** and not **Log x** or **Log(x)**.

(7) Improperly concatenating variables. For example, **xy** is a variable named xy, whereas **x y** or **x*y** means x *times* y.

(8) Using **y** instead of **y[x]** in **DSolve** and **NDSolve**.

(9) Omitting the **Evaluate** command when using **Table** inside a plotting command. See *The Evaluate Command* in this chapter.

(10) Using lower-case instead of upper-case letters in *Mathematica* commands. For example, *Mathematica* does not recognize **Findroot** or **findroot**; the correct command is **FindRoot**.

Error and Warning Messages

You should pay close attention to the error and warning messages that *Mathematica* generates. Although these messages may seem cryptic at first, you will find that they give valuable clues to locating mistakes in input. For example, suppose that we execute the following command without first defining a function $u(x)$.

```
In[47]:= Plot[u[x], {x, 0, 3}];

      Plot::plnr: u[x] is not a machine-size real number
         at x = 1.2499999999999977`*^-7.

      Plot::plnr: u[x] is not a machine-size real number
         at x = 0.121700974718747367`.

      Plot::plnr: u[x] is not a machine-size real number
         at x = 0.254426399578121076`.

   General::stop: Further output of Plot::plnr will
      be suppressed during this calculation.
```

This imposing error message is *Mathematica*'s way of telling you that whatever it is that you're trying to plot is not evaluating to a real number. In this case, we tried to plot a function that hadn't been defined, so of course it didn't evaluate to a real number.

It is important to distinguish between *error* messages and *warning* messages. An error message means that something is really wrong, and *Mathematica* will probably not be able to produce any output. A warning message is an indication that something might be wrong or that *Mathematica* can't do exactly what you've instructed it to do. Here are two examples:

```
In[48]:= sol[x_] = y[x] /. First[NDSolve[
           {y'[x] == y[x]^2, y[0] == 1}, y[x], {x, 0, 1}]]
```

    NDSolve::ndsz: At x == 0.999981747526144459`, step
        size is effectively zero; singularity suspected.

Out[48]= InterpolatingFunction[0., 0.999982, <>][x]

This warning message indicates that **NDSolve** was unable to generate a solution on the entire interval requested, and may occur, for example, if the solution has a vertical asymptote.

```
In[49]:= sol[1]
```

    InterpolatingFunction::dmval:
     Input value {1} lies outside the range of data in
        the interpolating function. Extrapolation will
        be used.

Out[49]= 7.42304×10^{89}

This second message occurred when we tried to evaluate the solution generated by **NDSolve** outside of the interval on which the solution is defined. In this case, the solution is not defined past 0.999982. Nevertheless, *Mathematica* extrapolates the solution and returns a value. This can be very misleading, especially if the actual solution blows up, which is why *Mathematica* issues a warning.

As you become comfortable using *Mathematica*, you may occasionally want to use the **Off** command to turn off certain warning messages. For example, you can turn off the warning message "NDSolve::ndsz" by typing **Off[NDSolve::ndsz]**. To turn it on again, enter **On[NDSolve::ndsz]**.

Recovering from Errors

From time to time you may experience certain problems using *Mathematica*. This is an unavoidable aspect of any complicated software product. We have tried to alert you to the most common and frustrating problems. More suggestions appear in the *Online Help* sections of Chapters 2–4; *Aborting Calculations* in Chapter 3; and

Quitting the Kernel in Chapter 4. If your computer is hung up and won't respond, type "ALT+." (on the *Macintosh*, "COMMAND+.", and in the *X Window System*, "Mod1+.") to abort the calculation. If all else fails, reboot the computer. Above all, remember to save your work often.

Problem Set C

Numerical Solutions
of Differential Equations

In this problem set you will use **NDSolve** and **Plot** to solve numerically and plot solutions to ordinary differential equations. These commands are explained in Chapters 3, 7, and 8. The solution to Problem 3 appears in the *Sample Notebook Solutions*.

1. We are interested in the solution $y = \phi(x)$ to the initial value problem (IVP)

$$y' = x^2 + y^2, \qquad y(0) = 1. \tag{i}$$

Observe that on the interval $0 \le x \le 1$, we have the inequalities

$$y^2 \le x^2 + y^2 \le 1 + y^2.$$

Therefore we must have $\phi_0(x) \le \phi(x) \le \phi_1(x)$, where ϕ_0, ϕ_1 are the solutions to the respective IVPs

$$\begin{cases} y' = y^2 \\ y(0) = 1, \end{cases} \qquad \begin{cases} y' = 1 + y^2 \\ y(0) = 1. \end{cases}$$

Solve these two IVPs explicitly and conclude that $\phi(x) \to \infty$ as $x \to x^*$, for some $x^* \in [\pi/4, 1]$. Then compute a numerical solution of (i), and find an approximate value of x^*, using **NDSolve**.

2. We are interested in the solution $y = \phi(x)$ to the initial value problem (IVP)

$$y' = y^2 - x, \qquad y(0) = 2. \tag{i}$$

Observe that on the interval $0 \le x \le 1$, we have the inequalities

$$y^2 \ge y^2 - x \ge y^2 - 1.$$

Therefore we must have $\phi_0(x) \ge \phi(x) \ge \phi_1(x)$, where ϕ_0, ϕ_1 are the solutions to the respective IVPs

$$\begin{cases} y' = y^2 \\ y(0) = 2, \end{cases} \qquad \begin{cases} y' = y^2 - 1 \\ y(0) = 2. \end{cases}$$

Solve these two IVPs explicitly and conclude that $\phi(x) \to \infty$ as $x \to x^*$, for some $x^* \in [0.5, 0.5 \ln 3]$. Then compute a numerical solution of (i), and find an approximate value of x^*, using **NDSolve**.

3. We shall study solutions $y = \phi_b(x)$ to the IVP

$$y' = (y - x)(1 - y^3), \qquad y(0) = b$$

for nonnegative values of x.

(a) Plot numerical solutions $\phi_b(x)$ for several values of b. Make sure to include values of b that are less than or equal to 0, between 0 and 1, equal to 1, and greater than 1.

(b) Now, on the basis of these plots, describe the behavior of the solution curves $\phi_b(x)$ for positive x, when $b \leq 0, 0 < b < 1, b = 1$, and $b > 1$. Identify limiting behavior and indicate where the solutions are increasing or decreasing.

(c) Finally, by combining your plots with a plot of the line $y = x$, show that the solution curves for $b > 1$ are asymptotic to this line. Explain from the differential equation why that is plausible. Plot the direction field of the differential equation to confirm your analysis.

4. We shall study solutions $y = \phi_b(x)$ to the IVP

$$y' = (y - x^2)(1 - y^2), \qquad y(0) = b$$

for nonnegative values of x.

(a) Plot numerical solutions $\phi_b(x)$ for several values of b. Make sure to include values of b that are less than -1, equal to -1, between -1 and 1, equal to 1, and greater than 1.

(b) Now, based on these plots, describe the behavior of the solution curves $\phi_b(x)$ for positive x, when $b < -1$, $b = -1$, $-1 < b < 1$, $b = 1$, and $b > 1$. Identify limiting behavior and indicate where the solutions are increasing or decreasing.

(c) Finally, by combining your plots with a plot of the parabola $y = x^2$, show that the solution curves for $b > 1$ are asymptotic to this parabola. Explain from the differential equation why that is plausible. Plot the direction field of the differential equation to confirm your analysis.

5. In this problem, we analyze the Gompertz-threshold model from Problem 13 of Problem Set B. That is, consider the differential equation

$$y' = y(1 - \ln y)(y - 3).$$

Using various nonnegative values for $y(0)$, find and plot several numerical solutions on the interval $0 \leq t \leq 10$. (The option **PlotRange** may be helpful here.) By examining the differential equation and analyzing your plots, identify all equilibrium solutions and discuss their stability.

6. Consider the IVP

$$\frac{dy}{dx} = \frac{x - e^{-x}}{y + e^y}, \qquad y(1.5) = 0.5.$$

(a) Use **NDSolve** to find approximate values of the solution at $x = 0, 1, 1.8$, and 2.1. Then plot the solution.

(b) If you did Problem 4 in Problem Set B, enter those results into this notebook. Compare the values of the actual solution and the numerical solution at the four specified points. Plot the actual solution and the numerical solution on the same graph.

(c) Now plot the numerical solution on several large intervals (*e.g.*, $0 \le x \le 10$ or $0 \le x \le 100$). Make a guess about the nature of the solution as $x \to \infty$. Try to justify your guess on the basis of the differential equation.

7. Consider the IVP

$$e^x \sin y - 2y \sin x + (e^x \cos y + 2 \cos x) \frac{dy}{dx} = 0, \qquad y(0) = 0.5.$$

(a) Use **NDSolve** to find approximate values of the solution at $x = -1, 1$, and 2. Then plot the solution.

(b) If you did Problem 5 in Problem Set B, enter those results into this notebook. Compare the values of the actual solution and the numerical solution at the three specified points. Plot the actual solution and the numerical solution on the same graph.

(c) Now plot the numerical solution on several large intervals (*e.g.*, $0 \le x \le 10$ or $0 \le x \le 100$). Make a guess about the nature of the solution as $x \to \infty$. Try to justify your guess on the basis of the differential equation.

8. We know that the differential equations

$$y' = e^{-x^2}, \qquad y' = e^{x^2}$$

cannot be solved in terms of elementary functions.

(a) Use **DSolve** to solve these two equations.

(b) Note the occurrence of the built-in functions **Erf** and **Erfi**. Mathematics texts refer to **Erf** as the *error function*. The function **Erfi** is called the *imaginary error function* and is defined by $\operatorname{erfi}(x) = -i \cdot \operatorname{erf}(ix)$, where $i = \sqrt{-1}$. Use the differentiation operator **D** to see that

$$\frac{d}{dx}(\operatorname{erf}(x)) = \frac{2}{\sqrt{\pi}} e^{-x^2} \quad \text{and} \quad \frac{d}{dx}(\operatorname{erfi}(x)) = \frac{2}{\sqrt{\pi}} e^{x^2}.$$

In fact,

$$\operatorname{erf}(x) = \frac{2}{\sqrt{\pi}} \int_0^x e^{-t^2}\, dt \quad \text{and} \quad \operatorname{erfi}(x) = \frac{2}{\sqrt{\pi}} \int_0^x e^{t^2}\, dt.$$

(c) Although one does not have elementary formulas for these functions, the numerical capabilities of *Mathematica* mean that we "know" these functions as well as we "know" elementary functions like $\tan x$. To illustrate this, evaluate $\text{erf}(x)$ at $x = 0, 1$, and 10.5, and plot $\text{erf}(x)$ on $-10 \le x \le 10$.

(d) Compute $\lim_{x \to \infty} \text{erf}(x)$ and $\int_{-\infty}^{\infty} e^{-t^2}\, dt$.

(e) Next solve the IVP

$$y' = 1 - 2xy, \qquad y(0) = 0$$

using **DSolve**. What does the solution do for large x? Find the value x at which the solution takes its maximum. What is the maximum value?

(f) Compute $\partial f / \partial y$ for the function $f(x, y) = 1 - 2xy$. Discuss stability of the equation in (e). Plot solutions on the intervals $[-3, 0]$ and $[0, 3]$, using several different initial conditions at the left endpoints of the intervals. Explain how these plots illustrate your conclusions about stability.

9. Consider the IVP

$$xy' + (\sin x)y = 0, \qquad y(0) = 1.$$

(a) Use **DSolve** to solve the IVP.

(b) Note the occurrence of the built-in function **SinIntegral**, called the *Sine Integral* function. Check that this function is an antiderivative of $\sin x / x$ by differentiating it.

(c) Evaluate $\lim_{x \to \infty} \text{Si}(x)$; you may have to do this indirectly. Plot $\text{Si}(x)$, and discuss the features of the graph.

(d) Do the same with the solution to the IVP.

(e) Now solve the IVP using **NDSolve**, and plot the computed solution using **Plot**.

You will find that *Mathematica* fails to deal with $\sin x / x$ at $x = 0$, even though the singularity is removable. One way to get around this is to give the initial condition $y = 1$ at a value of x extremely close to, but not equal to, zero. Since we are only finding an approximate solution with **NDSolve** anyhow, there shouldn't be much harm done as long as the amount by which we move the initial condition is small compared with the error we expect from the numerical procedure. Alternatively, you could try redefining the differential equation to be $y' + f(x)y = 0$ with

$$f(x) = \begin{cases} \sin x / x, & x \ne 0 \\ 1, & x = 0. \end{cases}$$

(f) Discuss the stability of the differential equation. Illustrate your conclusions by graphing solutions with different initial values on the interval $[-10, 10]$.

10. Solve the following initial value problems numerically, then plot the solutions. On the basis of your plots, predict what happens to each solution as x increases. In particular, if there is a limiting value for y, either finite or infinite, then find it. If it is unclear from the plot you've made, try replotting on a larger interval. Another possibility is that the solution blows up in finite time. If so, estimate the time.

   (a) $y' = e^{-2x} + \dfrac{1}{1 + y^2}$, $\quad y(0) = -1$

   (b) $y' = e^{-x} + y^2$, $\quad y(0) = -2$

   (c) $y' = \cos x - y^3$, $\quad y(0) = 0$

   (d) $y' = (\sin x)y - y^2$, $\quad y(0) = 1$.

11. Solve the following initial value problems numerically, then plot the solutions. On the basis of your plots, predict what happens to each solution as x increases. See Problem 10 for more instructions.

   (a) $y' = 5x - 3\sqrt{y}$, $\quad y(0) = 2$

   (b) $y' = (x^2 - y^2) \sin y$, $\quad y(0) = -1$

   (c) $y' = \dfrac{y^2 + 2xy}{x^2 + 3}$, $\quad y(1) = 2$

   (d) $y' = -2x + e^{-xy}$, $\quad y(0) = 1$.

12. This problem illustrates one of the possible pitfalls of blindly applying numerical methods without paying attention to the theoretical aspects of the differential equation itself. Consider the equation

$$-4\,x^2 + 2\,y(x) + x\,y'(x) = 0.$$

   (a) Use the *Mathematica* program in Chapter 7 to compute the Euler Method approximation to the solution with initial condition $y(-0.5) = 4.25$, using step size $h = 0.2$ and $n = 10$ steps. The program will generate a list of ordered pairs of numbers representing the (x, y) coordinates of points of the approximate solution. Use **ListPlot** with the option **PlotJoined -> True** to obtain a piecewise linear graph of the approximate solution.

   (b) Now repeat part (a) with step size $h = 0.1$ and $n = 20$ steps. Compare the values of the two approximations at $x = 0.1$.

   (c) Next try to use **NDSolve** to find an approximate solution on the interval $(-0.5, 1.5)$, and plot it with **Plot**. What is the interval on which the approximate solution is defined?

   (d) Solve the equation exactly and graph the solutions for the initial conditions $y(0) = 0$, $y(-0.5) = 4.25$, $y(0.5) = 4.25$, $y(-0.5) = -3.75$, and $y(0.5) = -3.75$. Now explain your results in (a)–(c). Could we have known, without solving the equation, whether to expect meaningful results in parts (a) and (b)? Why? Can you explain how **NDSolve** avoids making the same "mistake"?

13. Consider the IVP

$$\frac{dy}{dx} = e^{-x} - 2y, \qquad y(-1) = 0.$$

(a) Use the *Mathematica* program in Chapter 7 to compute the Euler Method approximation to $y(x)$ with step size $h = 0.5$ and $n = 4$ steps. The program will generate a list of ordered pairs (x_i, y_i). Use **ListPlot** with the option **PlotJoined -> True** to display the piecewise linear function connecting the points (x_i, y_i). Repeat with $h = 0.2$ and $n = 10$.

(b) Find the exact solution of the IVP and plot the exact solution and the two Euler approximations on the same graph. Label the three curves.

(c) Now use the Euler Method program with $h = 0.5$ to approximate the solution on the interval $[-1, 9]$. Plot both the approximate and exact solutions on this interval. How close is the approximation to the exact solution as x increases? In light of the discussion of stability in Chapters 5 and 7, explain your results in parts (a) and (b).

14. Consider the IVP

$$\frac{dy}{dx} = y - 4e^{-3x}, \qquad y(0) = 1.$$

(a) Use the *Mathematica* program in Chapter 7 to compute the Euler Method approximation to $y(x)$ with step size $h = 0.5$ and $n = 6$ steps. The program will generate a list of ordered pairs (x_i, y_i). Use **ListPlot** with the option **PlotJoined -> True** to display the piecewise linear function connecting the points (x_i, y_i). What appears to be happening to y as x increases?

(b) Repeat part (a) with $h = 0.2$ and $n = 15$, and then with $h = 0.1$ and $n = 30$. How are the approximate solutions changing as the step size decreases? Can you make a reliable prediction about the long-term behavior of the solution?

(c) Use **NDSolve** to find an approximate solution and plot it on the interval $[0, 3]$. Now what does it look like y is doing as x increases? Next plot the solution from **NDSolve** on a larger interval (going at least to $x = 10$). Again, what is happening to y as x increases?

(d) Solve the IVP exactly and compare the exact solution to the approximations found above. In light of the discussion of stability in Chapters 5 and 7, explain your results in parts (a)–(c).

15. Consider the IVP

$$\frac{dy}{dx} = 2y - 2 + 3e^{-x}, \qquad y(0) = 0.$$

(a) Use the *Mathematica* program in Chapter 7 to compute the Euler Method approximation to $y(x)$ with step size $h = 0.2$ and $n = 10$ steps. The program will generate a list of ordered pairs (x_i, y_i). Use **ListPlot** with the option **PlotJoined -> True** to display the piecewise linear function connecting the points (x_i, y_i) What appears to be happening to y as x increases?

(b) Repeat part (a) with $h = 0.1$ and $n = 20$, and then with $h = 0.05$ and $n = 40$. How are the approximate solutions changing as the step size decreases? Can you make a reliable prediction about the long-term behavior of the solution?

(c) Use **NDSolve** to find an approximate solution and plot it on the interval $[0, 2]$. Now what does it look like y is doing as x increases? Next plot the solution from **NDSolve** on a larger interval (going at least to $x = 5$). Again, what is happening to y as x increases?

(d) Solve the IVP exactly and compare the exact solution to the approximations found above. In light of the discussion of stability in Chapters 5 and 7, explain your results in parts (a)–(c).

16. We know that e^x is the solution to the IVP

$$\frac{dy}{dx} = y, \qquad y(0) = 1,$$

so if we solve this IVP numerically we get approximate values for e^x. Use **NDSolve**, employing the accuracy options discussed in Chapter 7, to calculate values for $e^{0.1}, e^{0.2}, \ldots, e$ that have 10 correct digits. Present your results in a table. In a second column print the values of e^x for $x = 0.1, 0.2, \ldots, 1$, obtained by using the built-in function **Exp**. Show at least 10 digits. Compare the two columns of values.

Chapter 9

Qualitative Theory of Second Order Linear Equations

Newton's second law of dynamics—force is equal to mass times acceleration—tells physicists that, in order to understand how the world works, they must pay attention to the forces. Since acceleration is a second derivative, the law also tells us that second order differential equations are likely to appear when we apply mathematics to study the real world.

The simplest second order differential equations are linear equations with *constant coefficients*:

$$ay'' + by' + cy = g(x).$$

These equations model a wide variety of physical situations, including oscillating springs, simple electric circuits, and the vibrations of tuning forks to produce sound and of electrons to produce light. In other situations, such as the motion of a pendulum, we may be able to approximate the resulting differential equation reasonably well by a linear differential equation with constant coefficients. Fortunately, we know (and *Mathematica* can apply) several techniques for finding explicit solution formulas to linear differential equations with constant coefficients.

Unfortunately, we often cannot find solution formulas for more general second order equations, even for linear equations with *variable coefficients*:

$$y'' + p(x)y' + q(x)y = g(x).$$

Such equations have important applications to physics. For example, Airy's equation,

$$y'' - xy = 0, \tag{1}$$

arises in diffraction problems in optics, and Bessel's equation,

$$y'' + \frac{1}{x}y' + y = 0, \tag{2}$$

occurs in the study of vibrations of a circular membrane and of water waves with circular symmetry.

When studying first order differential equations for which exact solution formulas were unavailable, we could turn to several other methods. By specifying an initial value, we could compute an approximate numerical solution. By letting the initial values vary, we could plot a one-parameter family of approximate solution curves and get a feel for the behavior of a general solution. Alternatively, we could plot the direction field of the differential equation and use it to draw conclusions about the qualitative behavior of solutions.

For second order equations, these methods are more cumbersome; you must specify two conditions to pick out a unique solution. We usually specify initial values for the function and its first derivative at some point. Then we can use numerical techniques to compute an approximate solution. In order to graph enough solutions to get an idea of their general behavior, we must construct a two-parameter family of solutions. Since the initial value does not determine the initial slope, we cannot draw a direction field for a second order equation.

In some applications, the differential equation is accompanied not by initial conditions, but by *boundary conditions*; *i.e.*, we specify values $y(x_0) = y_0$ and $y(x_1) = y_1$ at two distinct points. The resulting problem is called a *boundary value problem*. In this situation, we cannot compute a numerical solution directly. Instead, we must search among all the solutions that satisfy one of the boundary conditions in order to find one that satisfies the second boundary condition.

In this chapter, we describe how to use *Mathematica* to find exact solutions of second order linear differential equations with constant coefficients. We also describe how to find and plot numerical solutions to more general second order differential equations. In addition, we describe a method for solving boundary value problems using *Mathematica*'s numerical solver. Finally, we introduce comparison methods and a more sophisticated geometric method for analyzing second order linear equations with variable coefficients. These qualitative methods have an advantage over the more obvious numerical and graphical methods. They more effectively yield information about properties shared by all solutions of a differential equation, about the oscillatory nature of solutions of an equation, and about the precise rate of decay or growth of solutions.

Comparison methods provide information on the solutions of a variable coefficient equation by comparing the equation with an appropriate constant coefficient equation. This possibility is suggested by the following example. For large x, the coefficient $1/x$ in Bessel's equation (2) is close to 0. So, Bessel's equation is close to the equation $y'' + y = 0$, whose general solution can be written as $y = R\cos(x - \delta)$. Thus, one might expect solutions to Bessel's equation to oscillate and look roughly like sine waves for large x. We present a result that validates such comparisons.

The geometric method is based on direction fields, which are not directly applicable to a second order equation. We show, however, how to construct a related first order equation whose direction field yields information about the solutions of the second order equation.

Second Order Equations with *Mathematica*

Mathematica's usual tools for finding symbolic solutions of differential equations work perfectly well for second order linear differential equations with constant coefficients. The syntax of the commmands is similar, except that we must specify more initial conditions.

EXAMPLE 1. Consider the differential equation

$$y'' + y' - 6y = 20e^x.$$

If we type

```
ode1 = y´´[x] + y´[x] - 6y[x] == 20Exp[x];
DSolve[ode1, y[x], x]
```

we get the solution

$$\{\{y[x] \to -5E^x + E^{-3x}C[1] + E^{2x}C[2]\}\}.$$

As expected, the general solution depends on two arbitrary constants. To solve the differential equation with initial conditions $y(0) = 0$ and $y'(0) = 1$, we would type

```
DSolve[{ode1, y[0] == 0, y´[0] == 1}, y[x], x] //Simplify
```

The specific solution with these initial values is

$$\left\{\left\{y[x] \to \frac{4E^{-3x}}{5} - 5E^x + \frac{21E^{2x}}{5}\right\}\right\}.$$

Mathematica can also solve certain boundary value problems. Let's continue using the same differential equation for our example, but take the boundary conditions $y(0) = 0$ at the first point and $y(\ln(2)) = 10$ at the second point. If we type

```
DSolve[{ode1, y[0] == 0, y[Log[2]] == 10}, y[x], x]
 //Simplify
```

we get the answer

$$\{\{y[x] \to 5E^x(-1 + E^x)\}\}.$$

EXAMPLE 2. Now consider the differential equation

$$y'' + \frac{1}{2}x^4y' + y + xy^3 = 0,$$

which is a nonlinear second order differential equation. *Mathematica* is unable to find an explicit formula for the general solution. The most straightforward method for understanding the behavior of a general solution of the equation is to plot numerical solutions corresponding to a wide range of initial values and initial slopes. Here are the *Mathematica* commands to do this.

```
ode2 = y´´[x] + x^4*y´[x]/2 + y[x] + x*y[x]^3 == 0
sol2[x_, y0_, yp0_] := y[x] /. First[NDSolve[
 {ode2, y[0] == y0, y´[0] == yp0}, y[x], {x, -1.2, 3}]]
Plot[Evaluate[Table[sol2[x, y0, yp0], {y0, -2, 2},
 {yp0, -1, 1, 0.5}]], {x, -1.2, 3}];
```

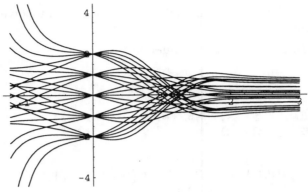

Figure 1

The result is shown in Figure 1. We see that the solutions seem to level off for positive x and seem to blow up for negative x.

Could we have obtained this qualitative information directly from the differential equation, instead of from the graph of a few solutions? One approach is to look at the solutions of differential equations that are similar to this one. For example, when x is small, the terms involving x are close to zero. So the solutions might be close to the solutions of the differential equation

$$y'' + y = 0.$$

We know that the solutions to this equation are sine curves. For example, the solution satisfying the initial conditions $y(0) = 0$, $y'(0) = 1$ is just $\sin(x)$. So, one might expect the solutions of this initial value problem to look like $\sin(x)$ for small values of x. In Figure 2, we have plotted $\sin(x)$ and the numerical solution to the initial value problem on the same axes.

The graph of the numerical solution in Figure 2 appears to level off as x increases. Looking again at the differential equation, we see that when x is large, the $x^4 y'$ term should dominate the other terms. So, the solutions to the differential equation should be close to those of

$$y'' + \frac{1}{2} x^4 y' = 0.$$

This equation can almost be solved explicitly; its general solution is given by

$$y(x) = A + B \int_0^x e^{-u^5/10} \, du.$$

The exponential function being integrated is nearly zero for large positive values of u. So, when x is large and positive, the solutions should become almost constant. When x is large and negative, the exponential term is large, and we expect the solutions to blow up as well.

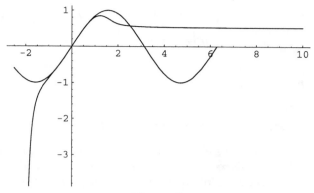

Figure 2

Finally, let's look at the boundary value problem

$$y'' + \frac{1}{2}x^4 y' + y + xy^3 = 0, \quad y(0) = 0, \quad y(1) = 1.$$

Here we have the same differential equation, but we have specified values for the solution curve at two different points. Although *Mathematica* can solve some boundary value problems numerically, it cannot solve this one. So, consider the following command, which tells *Mathematica* to compute a numerical solution as a function of the first derivative:

```
nsol2[s_] := NDSolve[{ode2, y[0] == 0, y'[0] == s},
  y[x], {x, 0, 1}]
trial[s_] := nsol2[s] /. x -> 1
```

Now when we evaluate **nsol2** at some real number, we get a function that can be used to compute numerical solutions to the differential equation. When we evaluate that procedure at $x = 1$, as is done by the **trial** command, we find the approximate value of that numerical solution at the desired endpoint in our initial value problem. After some trial and error, one finds in this case that **trial[1.2616]** produces the output

$$\{\{y[1] \rightarrow 1.00003\}\}$$

So, the solution to the boundary value problem is (approximately) the numerical solution with initial conditions $y(0) = 0$ and $y'(0) = 1.2616$. This technique is called the *shooting method*.

Comparison Methods

In this section, we discuss the Sturm Comparison Theorem. This theorem is a precise form of the rough comparisons we carried out in the previous section. In addition, we will discuss the relation between the zeros of two linearly independent solutions of a second order linear equation.

STURM COMPARISON THEOREM. *Let $u(x)$ be a solution to the equation*

$$y'' + q(x)y = 0,$$

and suppose $u(a) = u(b) = 0$ for some $a < b$ (but u is not identically zero). Let $v(x)$ be a solution to the equation

$$y'' + r(x)y = 0,$$

where $r(x) \geq q(x)$ for all $a \leq x \leq b$. Then $v(x) = 0$ for some x in $[a, b]$.

We will defer the proof of this theorem briefly, preferring instead to explain how to use it. We typically use this result to compare a variable coefficient equation with an appropriate constant coefficient equation. Consider, for example, the equation

$$y'' + \frac{1}{x}y = 0. \tag{3}$$

Let K be a positive number. If $0 < x \leq K$, then $1/x \geq 1/K$. We now apply the above theorem with $q(x) = 1/K$ and $r(x) = 1/x$. Thus, we compare equation (3) to the constant coefficient equation $y'' + (1/K)y = 0$, whose general solution is $u(x) = R\cos(\sqrt{1/K}x - \delta)$. Given any interval in $(0, K]$ of length $\pi\sqrt{K}$, the value of δ can be chosen so that $u(x) = 0$ at the endpoints of the interval. Then the Sturm Comparison Theorem implies that every solution of (3) has a zero on this interval. This argument implies that on $(0, K]$, the zeros of (3) cannot be farther apart than $\pi\sqrt{K}$.

In particular, solutions of (3) must oscillate as $x \to \infty$, though the oscillations may become less frequent as x increases. Indeed, by turning the above comparison around, one concludes that the zeros of solutions of (3) in $[K, \infty)$ must be at least $\pi\sqrt{K}$ apart.

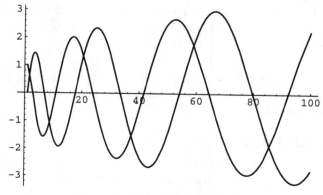

Figure 3

Figure 3 shows two representative solutions of (3) computed with **NDSolve**. As you can see, the predictions of the previous paragraph are confirmed by the graph.

The Sturm Comparison Theorem can also be used to study the zeros of Airy's equation and of Bessel's equation. In the latter case, a substitution is made to eliminate the y' term. See Problems 15 and 16 in Problem Set D.

The Interlacing of Zeros

You may have noticed that the curves in Figure 3 take turns crossing the horizontal axis. We say that the zeros of the two solutions are *interlaced*: between any two zeros of one solution there is a zero of the other. This is a common phenomenon— the property of interlaced zeros holds for any pair of linearly independent solutions of a homogeneous second order linear differential equation

$$y'' + p(x)y' + q(x)y = 0. \tag{4}$$

If $p(x) = 0$, as is the case in equation (1), the proof of the interlacing result follows from an application of the Sturm Comparison Theorem by comparing the equation with itself. For the more general equation (4), the proof is based on the Wronskian. Let $y_1(x)$ and $y_2(x)$ be any two linearly independent solutions of (4). Recall that the Wronskian of $y_1(x)$ and $y_2(x)$ is

$$W(x) = y_1(x)y_2'(x) - y_1'(x)y_2(x).$$

Since y_1 and y_2 are solutions of (4), it follows that W satisfies the differential equation

$$W' = -p(x)W,$$

whose general solution is

$$W(x) = Ce^{-\int p(x)\,dx}.$$

This formula shows that $W(x)$ does not change sign—it always has the same sign as the constant C, which must be nonzero (or else y_1 and y_2 would be linearly dependent).

Assume that $y_1(a) = y_1(b) = 0$ and that there are no zeros of y_1 between a and b. Then

$$W(a) = -y_1'(a)y_2(a),$$
$$W(b) = -y_1'(b)y_2(b).$$

Since y_1 does not change sign between a and b, we know $y_1'(a)$ and $y_1'(b)$ have opposite signs. Then, since $W(a)$ and $W(b)$ have the same sign, $y_2(a)$ and $y_2(b)$ must have opposite signs. Therefore y_2 must be zero somewhere between a and b. Similarly, y_1 must have a zero between any two zeros of y_2.

EXERCISE. The solutions to Airy's equation and Bessel's equation are considered so important that they have been given names in *Mathematica*. The Airy functions $Ai(x)$ and $Bi(x)$ are a fundamental set of solutions for (1), and the Bessel functions $J_0(x)$ and $Y_0(x)$ are a fundamental set of solutions for (2). These functions are built into *Mathematica* as **AiryAi[x]**, **AiryBi[x]**, **BesselJ[0, x]**, and

`BesselY[0, x]`. Use *Mathematica* to check that the zeros of the two solutions `AiryAi[x]` and `AiryBi[x]` to Airy's equation are interlaced. Do the same for the Bessel functions `BesselJ[0, x]` and `BesselY[0, x]`.

Proof of the Sturm Comparison Theorem

The proof is based on the Wronskian, as was the proof of the interlacing result. Recall the hypotheses: $u'' + q(x)u = 0$ and $v'' + r(x)v = 0$, with $q(x) \leq r(x)$ between two zeros a and b of u. Since u is not identically zero, we can assume u does not change sign between a and b. (If it did it would have a zero between a and b, and we could look at a smaller interval.) Assume, for instance, that $u > 0$ between a and b. Since $u = 0$ at a and b, it follows that $u'(a) > 0$ and $u'(b) < 0$. (Notice that these derivatives cannot be zero because then by the uniqueness theorem u would be identically zero.) We want to prove that v is zero somewhere in $[a, b]$; we suppose it is not, say $v(x) > 0$ throughout $[a, b]$, and argue to obtain a contradiction.

Let $W(x)$ be the Wronskian of u and v,

$$W(x) = u(x)v'(x) - u'(x)v(x);$$

then

$$W(a) = -u'(a)v(a) < 0,$$
$$W(b) = -u'(b)v(b) > 0.$$

Also, since $q(x) \leq r(x)$,

$$W' = uv'' - u''v = u(-r(x)v) - (-q(x)u)v = (q(x) - r(x))uv \leq 0.$$

But this is impossible—the last inequality implies W does not increase between a and b, yet $W(a) < 0 < W(b)$. This contradiction means that v cannot have the same sign throughout $[a, b]$, and therefore v must be zero somewhere in the interval.

A Geometric Method

As mentioned above, direction fields are not directly applicable to second order equations. We now show, however, that with a given second order homogeneous equation, we can associate a first order equation whose direction field yields information about the solutions of the second order equation. (Our approach is similar to the classical method of associating a first order Riccati equation to a second order linear equation. The substitution we use is akin to the Prüfer substitution for Sturm-Liouville systems; see G. Birkhoff and G.-C. Rota, **Ordinary Differential Equations**, 3rd ed., J. Wiley and Sons, Inc., 1978.)

Consider the homogeneous equation

$$y'' + p(x)y' + q(x)y = 0. \tag{5}$$

Let

$$z = \arctan\left(\frac{y}{y'}\right);$$

then

$$z' = \left(1 + \left(\frac{y}{y'}\right)^2\right)^{-1} \frac{d}{dx}\left(\frac{y}{y'}\right) = \frac{y'^2}{y'^2 + y^2} \frac{y'^2 - yy''}{y'^2} = \frac{y'^2 - yy''}{y'^2 + y^2}.$$

Also, since $\tan z = y/y'$, notice that

$$\sin z = \frac{y}{\sqrt{y'^2 + y^2}}, \qquad \cos z = \frac{y'}{\sqrt{y'^2 + y^2}}.$$

Then if y satisfies (5) and y is not identically zero, it follows that

$$z' = \frac{y'^2 - yy''}{y'^2 + y^2} = \frac{y'^2 - y(-p(x)y' - q(x)y)}{y'^2 + y^2} = \frac{y'^2 + p(x)yy' + q(x)y^2}{y'^2 + y^2}.$$

In other words,

$$z' = \cos^2 z + p(x)\sin z \cos z + q(x)\sin^2 z. \tag{6}$$

We have shown that for every solution y of (5) that is not identically zero, there is a corresponding solution z of (6). (This is not a one-to-one correspondence—every constant multiple of a solution y corresponds to the same solution z.) Although the solution curves of (6) will be different from the solution curves of (5), we now show that we can learn about the solutions of (5) by studying the solutions of (6), specifically, by considering the direction field of (6).

The Constant Coefficient Case

We begin by considering the equation

$$y'' - y = 0,$$

whose general solution is

$$y = c_1 e^x + c_2 e^{-x}.$$

The corresponding first order equation is

$$z' = \cos^2 z - \sin^2 z.$$

In Figure 4, we show the direction field of this equation, which we produced with the *Mathematica* command

```
<<Graphics`PlotField`
PlotVectorField[{1, Cos[z]^2 - Sin[z]^2}, {x, 0, 10},
   {z, -Pi/2, Pi/2}, PlotPoints -> 20, Axes -> True,
   Frame -> True, ScaleFunction -> (1&), AspectRatio -> 1]
```

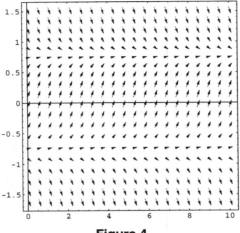

Figure 4

Notice that we plot z from $-\pi/2$ to $\pi/2$; this is the range of values taken on by the arctan function. You may wonder what happens if z goes off the bottom of the graph. The answer is that z "wraps around" to the top of the graph. Recall that $z = \arctan(y/y')$; the points $z = \pm\pi/2$ correspond to $y' = 0$, and when y' changes sign, then so does z by passing from $-\pi/2$ to $\pi/2$ (or vice versa).

Next, observe that there is an unstable equilibrium at $z = -\pi/4$, which corresponds to $y/y' = \tan(-\pi/4) = -1$; this represents solutions of the form $y = ce^{-x}$. Similarly, the stable equilibrium $z = \pi/4$ corresponds to $y/y' = \tan(\pi/4) = 1$, which represents the solutions $y = ce^x$. The fact that most solutions of the equation satisfied by z (those with initial condition other than $-\pi/4$) approach the stable equilibrium $z = \pi/4$ corresponds to the fact that most solutions $y = c_1 e^x + c_2 e^{-x}$ (those with $c_1 \neq 0$) grow like e^x as x increases.

Another basic example is
$$y'' + y = 0,$$
for which the corresponding first order equation is
$$z' = \cos^2 z + \sin^2 z = 1.$$

That is, z simply increases linearly, though every time it reaches $\pi/2$ it wraps around to $-\pi/2$. This corresponds to the oscillation of solutions y; for every time $z = \arctan(y/y')$ passes through zero, then so must y, and vice versa.

EXERCISE. Consider the general second order linear homogeneous equation with constant coefficients:
$$ay'' + by' + c = 0.$$

Investigate how the roots of the characteristic equation, if real, correspond to equilibrium solutions for z. Show that if the roots of the characteristic equation are complex, then z' is always positive.

The Variable Coefficient Case

The examples above suggest ways to draw parallels between solutions z of (6) and solutions y of (5). First of all, if z is positive, then y and y' have the same sign. This implies that y is moving away from zero as x increases. We say in this case that y is *growing*, meaning that $|y|$ is increasing. Similarly, if z is negative then y and y' have opposite signs, which implies that y is moving toward zero as x increases—we say in this case that y is *decaying*, meaning that $|y|$ is decreasing. If z changes sign by passing through zero, then so does y, and if z passes through $\pm\pi/2$, then y' changes sign, showing that y has passed through a local maximum or minimum.

In light of these observations, let us summarize what we can predict about the *long-term behavior* of a solution y of (5) in terms of the corresponding solution z of (6).

- If z remains positive as x increases, then y grows away from zero.
- If z remains negative as x increases, then y decays toward zero.
- If z continues to increase from $-\pi/2$ to $\pi/2$ and wraps around to $-\pi/2$ again, then y *oscillates* as x increases.

Notice from (6) that $z' = 1$ whenever $z = 0$, so it is not possible for z to decrease through zero. Thus the long-term behavior of z will usually fall into one of the three categories above.

We can be more precise about the growth or decay rate of y in cases where z approaches a limiting value. If the limiting value is θ, then y/y' approaches $\tan\theta$ as x increases. In other words, $y' \approx (\cot\theta)y$ for large x. Thus

$$y \approx ce^{(\cot\theta)x}$$

for large x, and $\cot\theta$ is the asymptotic exponential growth (or decay) rate of y. If z approaches zero as x increases, then y/y' approaches zero as well. One can show that y grows (if $z > 0$) or decays (if $z < 0$) faster than any exponential function. Similarly, if z approaches $\pi/2$, then y grows slower than exponentially, and if z approaches $-\pi/2$, then y decays slower than exponentially. Since y'/y approaches zero in these cases, y could grow or decay toward a finite, nonzero value.

Let us now see what the direction field of (6) tells us about solutions of Airy's and Bessel's equations.

Airy's Equation

For Airy's equation,

$$y'' - xy = 0, \tag{7}$$

the corresponding first order equation is

$$z' = \cos^2 z - x\sin^2 z.$$

Figure 5 shows the direction field of this equation, plotted by a *Mathematica* command similar to the one used for Figure 4.

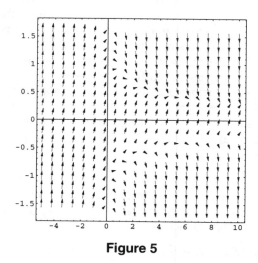

Figure 5

Notice that z is increasing steadily for negative x, so solutions of (7) must oscillate for negative x. For positive x, it is evident that solutions of (7) cannot oscillate; z passes through zero at most once, after which it remains positive. Hence the corresponding solution of (7) grows away from zero as x increases. It appears that once z becomes positive, it decreases to zero, indicating that most solutions of (7) grow faster than exponentially.

We now know there are solutions of (7) that grow very fast as x increases, and there are no oscillating solutions for positive x. Is it possible to have a decaying solution of (7)? Equivalently, is it possible for z to remain negative as x increases? Figure 5 strongly suggests that there is such a solution z, and hence that there is a solution y to (7) that decays toward zero at a rate faster than exponential.

The Airy function $\text{Ai}(x)$ is, by definition, the unique (up to a constant multiple) solution of (7) that decays as x increases. As expected, the other Airy function $\text{Bi}(x)$ grows as x increases. You can see what these functions look like by plotting **AiryAi[x]** and **AiryBi[x]** in *Mathematica*. The general solution of (7) is a linear combination of $\text{Ai}(x)$ and $\text{Bi}(x)$.

Bessel's Equation

For Bessel's equation

$$y'' + \frac{1}{x}y' + y = 0, \tag{8}$$

the corresponding first order equation is

$$z' = \cos^2 z + \frac{1}{x}\sin z \cos z + \sin^2 z.$$

Figure 6 shows the direction field of this equation.

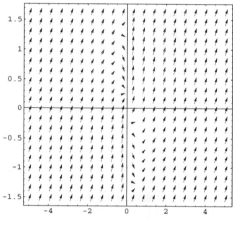

Figure 6

The picture is simpler than the one for Airy's equation. When x is away from zero, z is increasing steadily. So, solutions of (8) oscillate for both positive and negative x. Near $x = 0$, the direction field is irregular because of the $1/x$ term in the differential equation. We can't tell what happens from this picture, but there is reason to expect that solutions of (8) will not behave nicely near this singularity in the differential equation.

One question this approach cannot answer easily is whether the amplitude of oscillating solutions grows, decays, or remains steady as x increases. Since the coefficient of the y' term in (8) is positive, we can expect solutions of (8) to behave like a physical oscillator with damping—the amplitude of oscillations should decrease as x increases. Since the damping coefficient goes to zero as x increases, we cannot be sure (without a more refined analysis) whether the amplitude of solutions of (8) must decrease to zero. In fact it does, as you can check by plotting the Bessel functions **BesselJ[0, x]** and **BesselY[0, x]** in *Mathematica*.

Other Equations

Here are some other differential equations for which you can try this method.

- Bessel's equation of order n,

$$y'' + \frac{1}{x}y' + \left(1 - \frac{n^2}{x^2}\right)y = 0,$$

for various n (above we studied the case $n = 0$). You can check your predictions using the functions **BesselJ[n, x]** and **BesselY[n, x]**, which are built into *Mathematica*. These functions are considered in Problem 18 of Problem Set D.

- The modified Bessel equation of order n,

$$y'' + \frac{1}{x}y' - \left(1 + \frac{n^2}{x^2}\right)y = 0,$$

for various n. The functions **BesselI[n, x]** and **BesselK[n, x]** are a fundamental set of solutions for this equation.

- The parabolic cylinder equation

$$y'' + \left(n + \frac{1}{2} - \frac{1}{4}x^2\right)y = 0$$

for various n. This equation arises in quantum mechanics, but its solutions are not built into *Mathematica*; you can check your predictions against approximate solutions from **NDSolve** instead. This equation is considered in Problem 17 of Problem Set D.

Problem Set D

Second Order Equations

The solution to Problem 1 appears in the *Sample Notebook Solutions*.

1. Airy's equation is the linear second order homogeneous equation $y'' = xy$. Although it arises in a number of applications, including quantum mechanics, optics, and waves, it cannot be solved exactly by the standard symbolic methods. In order to analyze the solution curves, let us reason as follows.

   (a) For x close to zero, the equation resembles $y'' = 0$, which has general solution $y = c_1 x + c_2$. We refer to this as a "facsimile" solution. Graph a numerical solution to Airy's equation with initial conditions $y(0) = 0$, $y'(0) = 1$, and the facsimile solution (with the same initial data) on the interval $(-2, 2)$. How well do they match?

   (b) For $x \approx -K^2 \ll 0$, the equation resembles $y'' = -K^2 y$, and the corresponding facsimile solution is given by $y = c_1 \sin(Kx + c_2)$. Again using the initial conditions $y(0) = 0$, $y'(0) = 1$, plot a numerical solution of Airy's equation over the interval $(-18, -14)$. Using the value $K = 4$, try to find values of c_1 and c_2 so that the facsimile solution matches well with the actual solution. Why shouldn't we expect the initial conditions for Airy's equation to be the appropriate initial conditions for the facsimile solution?

   (c) For $x \approx K^2 \gg 0$, Airy's equation resembles $y'' = K^2 y$, which has solution $y = c_1 \sinh(Kx + c_2)$. (The hyperbolic sine function is called `Sinh[x]` in *Mathematica*.) Plot a numerical solution of Airy's equation together with a facsimile solution (with $K = 4$) on the interval $(14, 18)$. In analogy with part (b), you have to choose values for c_1 and c_2 in the facsimile solution. (Try $c_1 = 1$, $c_2 = 0$.)

   (d) Plot the numerical solution of Airy's equation on the interval $(-20, 2)$. What does the graph suggest about the frequency and amplitude of oscillations as $x \to -\infty$? Could any of that information have been predicted from the facsimile analysis?

2. Consider Bessel's equation of order zero

$$x^2 y'' + x y' + x^2 y = 0 \qquad (i)$$

with initial data $y(0) = 1$, $y'(0) = 0$. The solution $J_0(x)$ is usually called the Bessel function of order zero of the first kind. Strictly speaking, this equation has a singularity at $x = 0$. However, this is one instance of a solution to a linear equation that exists outside the expected domain of definition. Still, as you will see below, the presence of the singularity can affect your computations.

(a) For x close to 0, x^2 is very small compared to x. The equation (i) is therefore approximately $x y' = 0$. Solve this equation with the preceding initial data. What does this "facsimile" solution to the original problem suggest to you about the behavior of J_0 near the origin?

(b) For x large and positive, x is small compared to x^2; and so we may approximate (i) by the equation $x^2(y'' + y) = 0$. Solve this equation with the same initial data. What does this suggest about the nature of the function J_0 for large x?

(c) Still thinking of x as large and positive, rewrite (i) in the form

$$y'' + \frac{1}{x} y' + y = 0.$$

If $x \approx K \gg 0$, we might approximate the equation by the constant coefficient equation

$$y'' + \frac{1}{K} y' + y = 0.$$

Solve that equation and see what further information you obtain about J_0. (If need be, choose a specific value for K, say $K = 100$, and examine the solution in order to refine your conclusion from (b).) Now explain why J_0 must be an even function, and so make the same deduction for large negative x.

(d) Throw caution to the winds and try **DSolve** on the original problem. Explain the result.

(e) Have a crack at the IVP with **NDSolve**. Can you find the value of the solution at $x = 1$? Explain your results.

3. This and some of the following problems concern models for the motion of a pendulum, which consists of a weight attached to a rigid arm of length L that is free to pivot in a complete circle. Neglecting friction and air resistance, the angle $\theta(t)$ that the arm makes with the vertical direction satisfies the differential equation

$$\theta''(t) + \frac{g}{L} \sin(\theta(t)) = 0, \qquad (i)$$

where $g = 32.2 \text{ ft/sec}^2$ is the gravitational acceleration constant. We will assume the arm has length 32.2 ft and so replace (i) by the simpler form

$$\theta'' + \sin \theta = 0. \qquad (ii)$$

(Alternatively, one can rescale time, replacing t by $\sqrt{g/L}\,t$, to convert (i) to (ii).) For motions with small displacements (θ small), $\sin\theta \approx \theta$, and (ii) can be approximated by the linear equation

$$\theta'' + \theta = 0. \qquad (iii)$$

This equation has general solution $\theta(t) = A\cos(t-\delta)$, with *amplitude A* and *phase shift* δ. Since all the solutions to the linear approximation (iii) have period 2π, we expect that, for small displacements, the solutions to the true pendulum (ii) should have period close to 2π. One can investigate how the period depends on the amplitude A by plotting a numerical solution of equation (ii) using initial conditions $\theta(0) = A, \theta'(0) = 0$ on an appropriate interval for various A. Estimate, to within 0.1, the period of the pendulum for the amplitudes $A = 0.1$, $0.7, 1.5$, and 3.0. What is happening to the accuracy of the linear approximation as the initial displacement increases?

4. In this problem, we'll look at what the pendulum does for various initial velocities (*cf.* Problem 3). Define a *Mathematica* function of v that will numerically solve the differential equation $(3.ii)$ using initial conditions $\theta(0) = 0, \theta'(0) = v$. Do likewise for equation $(3.iii)$. Plot the solutions to both the nonlinear equation $(3.ii)$ and the linear equation $(3.iii)$ for each of the following values of the initial velocity v: $1, 1.99, 2, 2.01$. Use a time interval from $t = 0$ to $t = 40$. Clearly mark which is the nonlinear solution and which is the linear solution on each plot. Compare the linear and nonlinear behavior in each case. For the nonlinear equation, interpret what the graph indicates the pendulum is doing physically. Speculate as to the true motion.

5. In this problem, we'll investigate the effect of damping on the pendulum, using the model

$$\theta'' + b\theta' + \sin\theta = 0$$

(*cf.* Problem 3). Define a *Mathematica* function of b that will find a numerical solution of this differential equation with initial conditions $\theta(0) = 0, \theta'(0) = 4$. Then plot the solution from $t = 0$ to $t = 20$ for the values $b = 0.5, 1$, and 2. Interpret what is happening *physically* in each case, *i.e.*, describe explicitly what the graph says the pendulum is doing. Now do the same for the linear approximation

$$\theta'' + b\theta' + \theta = 0,$$

for the same values of b. Compare the linear and nonlinear behavior in each case.

6. In this problem, we'll look at the effect of a periodic external force on the pendulum, using the model

$$\theta'' + 0.05\theta' + \sin\theta = 0.3\cos\omega t$$

(*cf.* Problems 3–5). We have chosen a value for the damping coefficient that is more typical of air resistance than the values in the previous problem. Write a *Mathematica* function of ω that numerically solves this equation with initial conditions $\theta(0) = 0, \theta'(0) = 0$. Then plot the result from $t = 0$ to $t = 60$ for the following values of the frequency ω: $0.6, 0.8, 1, 1.2$. Which frequency moves the pendulum farthest away from its equilibrium position? Do the same analysis for the linear approximation

$$\theta'' + 0.05\theta' + \theta = 0.3\cos\omega t$$

and answer the same question. For which frequencies do the linear and nonlinear equations have widely different behaviors? Which forcing frequency seems to induce resonance-type behavior in the pendulum? Graph that solution on a longer interval and decide whether the amplitude goes to infinity.

7. In many applications, second order equations come with *boundary conditions* rather than *initial conditions*. For example, consider a cable that is attached at each end to a post, with both ends at the same height (let us call this height $y = 0$). If the posts are located at $x = 0$ and $x = 1$, the height y of the cable as a function of x satisfies the differential equation

$$y'' = c\sqrt{1 + (y')^2}$$

with boundary conditions $y(0) = 0, y(1) = 0$. The constant c depends on the length of the cable; for this problem we'll use $c = 1$. *Mathematica* can't quite solve this boundary problem. Instead, use the shooting method to find the value of $y'(0)$ that leads to a solution satisfying the condition $y(1) = 0$. Graph the solution on $[0, 1]$ and determine the maximum dip in the cable.

8. The problem of finding the function $u(x)$ satisfying

$$\begin{cases} a(x)u''(x) + a'(x)u'(x) = f(x), 0 \le x \le 1 \\ u(0) = u(1) = 0 \end{cases}$$

arises in studying the longitudinal displacements in a longitudinally loaded elastic bar. The bar is of length 1, its left end is at $x = 0$, and its right end at $x = 1$. In the differential equation above, $f(x)$ represents the external force on the bar (which is assumed to be longitudinal, *i.e.*, directed along the bar), $a(x)$ represents both the elastic properties and the cross-sectional area of the bar, and $u(x)$ is the longitudinal displacement of the bar at the point x. This problem is an example of a *boundary value problem*; the conditions $u(0) = u(1) = 0$, which specify that the ends of the bar are fixed, are called boundary conditions. The function $a(x)$ may be constant; this is the case if neither the elastic properties nor the cross-sectional area depends on the position x in the rod, *i.e.*, if the rod is uniform. But if the rod is not uniform, then $a(x)$ is not constant, and the equation has variable coefficients. Similarly, $f(x)$ will be constant only if the external force is applied uniformly along the rod.

(a) Take $a(x) = 1 + x^2$ and $f(x) = 50x^4$. Using either **DSolve** or **NDSolve**, find a solution $u(x)$ to the boundary problem and plot it on the interval $[0, 1]$.

(b) Now suppose $a(x) = 1 + \tanh(x)$ and $f(x) = 50x^4$. *Mathematica* takes an extremely long time to solve this boundary problem numerically. Instead, use the shooting method to find the value of $u'(0)$ that leads to a solution satisfying the condition $u(1) = 0$. Graph the solution on $[0, 1]$. What is the maximum displacement, and where does it occur?

9. In this problem, we study the effects of air resistance.

(a) A paratrooper steps out of an airplane at a height of 1000 ft and after 5 seconds opens her parachute. Her weight, with equipment, is 195 lbs. Let $y(t)$ denote her height above the ground after t seconds. Assume that the air resistance is $0.005y'(t)^2$ lbs in free fall and $0.6y'(t)^2$ lbs with the chute open. At what height does the chute open? How long does it take to reach the ground? At what velocity does she hit the ground? (This model assumes that air resistance is proportional to the *square* of the velocity and that the parachute opens instantaneously.)

(*Hint:* Pay attention to units. Recall that the mass of the paratrooper is $195/32$, measured in lb sec^2/ft. Here, 32 is the acceleration due to gravity, measured in ft/sec^2.)

(b) Let $v = y'$ be the velocity during the second phase of the fall (while the chute is open). One can view the equation of motion as an autonomous first order ODE in the velocity:

$$v' = -32 + \frac{192}{1950}v^2.$$

Make a qualitative analysis of this first order equation, finding in particular the critical or equilibrium velocity. This velocity is called the terminal velocity. How does the terminal velocity compare with the velocity at the time the chute opens and with the velocity at impact?

(c) Assume the paratrooper is safe if she strikes the ground at a velocity within 5% of the terminal velocity in (b). Except for the initial height, use the parameters in (a). What is the lowest height from which she may parachute safely? (*Please do not try this at home!*)

10. This problem is based on Problems 25 and 26 in Boyce & DiPrima, Section 3.8. Consider the IVP

$$u'' + cu' + u = 0, \quad u(0) = 2, \quad u'(0) = 0.$$

(a) Find the solution. For $c = 0.25$, graph the solution. Determine an approximate time τ at which the solution becomes "negligible". Interpret negligible to mean that no meaningful difference between the curve and the x-axis can be discerned on your plot. Repeat this process for the values $c = 0.25, 0.5, \ldots, 1.75, 2$.

Make a list of values (c, τ) corresponding to the eight values of c. Plot the resulting points. (Use **ListPlot** and the **PlotJoined** $\rightarrow$ **True** option.) How does τ depend on c? What does the theory of damped vibrations predict about the relationship between τ and c?

(b) Now plot the solution curves for the values $c = 1.75, 1.8, \ldots, 2.2, 2.25$. The issue we would like to resolve is: at what point does the transition from oscillatory to nonoscillatory behavior occur? The characteristic equation tells you it is $c = 2$. Why? By adjusting your domain and range, try to verify that. If you cannot, explain what prevents you from doing so. (*Hint:* For $c = 1.9$, for example, compute where the curve first crosses the x-axis. What is the magnitude at nearby points?)

11. This problem is based on Problems 18 and 19 in Boyce & DiPrima, Section 3.9. Consider the IVP

$$u'' + u = 3\cos(\omega t), \quad u(0) = 0, \quad u'(0) = 0.$$

(a) Find the solution (using **DSolve**). For $\omega = 0.5, 0.6, 0.7, 0.8, 0.9$, plot the solution curves on the interval $0 \le x \le 10$. Note that $\omega_0 = 1$ is the natural frequency of the homogeneous equation. Describe how the solution curves behave as ω gets closer to 1.

(b) Notice that the formula in (a) is invalid when $\omega = 1$. Find and plot the solution curve for $\omega = 1$. What phenomenon is exhibited? Corroborate your conclusion by plotting on a longer interval.

(c) Compare the plots for $\omega = 1$ and $\omega = 0.9$ on a longer interval to see that the behavior for $\omega = 0.9$ is different from that in (b). What phenomenon is exhibited by the curve for $\omega = 0.9$?

12. In this problem we study how solutions of the initial value problem

$$\frac{d^2 y}{dt^2} + 0.15\frac{dy}{dt} - y + y^3 = 0, \quad y(0) = a, \quad y'(0) = 0$$

depend on the initial value a.

Plot a numerical solution of this equation from $t = 0$ to $t = 40$ for each of the initial values $a = 0.5, 1, 1.5, 2, 2.5$. You may have to adjust **MaxSteps** in **NDSolve** in order to make some of these plots.

Describe how increasing the initial value of y affects the solutions, both in terms of their limiting behavior and their general appearance. (In making comparisons, pay careful attention to the scale on the y-axis. Better still, use the **PlotRange** option to force the same scale on all of your plots.)

Finally, plot all five solutions on one graph. Would such a picture be possible for solutions of a first order differential equation? Why or why not?

13. In this problem, we consider the long-term behavior of solutions of the initial value problem

$$\frac{d^2y}{dt^2} + 0.15\frac{dy}{dt} - y + y^3 = 0.3\cos(\omega t), \quad y(0) = 0, \quad y'(0) = 0$$

for various frequencies ω in the forcing term.

Plot a numerical solution of this equation from $t = 0$ to $t = 100$ for each of the eight frequencies $\omega = 0.8, 0.9, \ldots, 1.4, 1.5$. You may have to adjust **MaxSteps** in **NDSolve** in order to make these plots. You should also expect *Mathematica* to take more time than usual in making them.

Describe and compare the different long-term behaviors you see. Due to the forcing term, all solutions will oscillate, but pay particular attention to the magnitude of the oscillations, and to whether or not there is a periodic pattern to them. Are there any similarities between your results for this nonlinear system and the phenomenon of resonance for linear systems with periodic forcing?

14. In this problem, we study the zeros of solutions of the second order differential equation

$$y'' + (2 + \sin x)y = 0. \tag{i}$$

(a) Using **NDSolve**, compute and plot several solutions of this equation with different initial conditions: $y(0) = c, y'(0) = d$. To be specific, choose three different values for the pair c, d and plot the corresponding solutions over $[0, 20]$. By inspecting your plots, find a number L that is an upper bound for the distance between successive zeros of the solutions. Then find a number l that is a lower bound for the distance between successive zeros.

(b) Information on the zeros of solutions of linear second order ODEs can be obtained from the Sturm Comparison Theorem (see Chapter 9). By comparing (i) with the equation $y'' + y = 0$, you should be able to get a value for L, and by comparing (i) with $y'' + 3y = 0$, you should be able to find l. How do these values compare with the values obtained in part (a)? (You will find it useful to note that the general solution of $y'' + ky = 0$, where k is a positive constant, can be written as $y = R\cos(\sqrt{k}x - \delta)$, with $R \geq 0, \delta$ arbitrary. R is called the *amplitude* and δ the *phase shift*.)

(c) Plot a solution of (i) and a solution of $y'' + y = 0$ on the same graph, and verify that between any two zeros of the latter solution, there is at least one zero of the solution of (i).

15. Consider Bessel's equation of order zero

$$x^2y'' + xy' + x^2y = 0.$$

(a) Using **NDSolve**, compute and plot several solutions of this equation with different initial conditions: $y(1) = c, y'(1) = d$. To be specific, choose three different values for the pair c, d and plot the corresponding solutions over $[0.1, 20]$. (Why isn't it a good idea to use $x = 0$ in the initial conditions?) By inspecting your plots, find a number L that is an upper bound for the distance between successive zeros of the solutions. Then find a number l that is a lower bound for the distance between successive zeros.

(b) Now confirm your findings with the Sturm Comparison Theorem. It doesn't apply directly, but if we introduce the new function $z(x) = x^{1/2}y(x)$, Bessel's equation becomes

$$z'' + (1 + 1/(4x^2))z = 0,$$

which has the form of the equation in the Comparison Theorem. Since y will have a zero wherever z has a zero, we can study the zeros of y by studying the zeros of z. By comparing the equation for z with the equation $z'' + z = 0$, determine an upper bound L on the distance between successive zeros of any solution of Bessel's equation.

(c) Let a be a positive number. For $x \in [a, \infty)$, the quantity $1 + 1/(4x^2)$ is less than or equal to the constant $1 + 1/(4a^2)$. By making an appropriate comparison, determine a lower bound l on the distance between successive zeros of any solution of Bessel's equation (for $x > a$). What is the limiting value of l as a goes to ∞? Approximately how far apart are the zeros when x is large? Did your graphical study lead to comparable values for l and L?

16. A solution of a second order linear differential equation is called *oscillatory* if its graph crosses the x-axis infinitely many times and *nonoscillatory* if it crosses only finitely many times. In this problem, we will be interested in determining the oscillatory nature of nonzero solutions to some second order linear ODEs. First consider the equation $y'' + ky = 0$, where k is a constant. If $k > 0$, it has the general solution $y = R\cos(\sqrt{k}x - \delta)$, with $R \geq 0, \delta$ arbitrary. R is called the *amplitude* and δ the *phase shift*. From this formula we see that any solution $y(x)$ has infinitely many zeros and hence crosses the axis infinitely many times, *i.e.*, is oscillatory. If $k < 0$, the general solution is $y(x) = c_1 e^{\sqrt{-k}x} + c_2 e^{-\sqrt{-k}x}$, so we see that solutions have at most one zero, *i.e.*, are nonoscillatory. If $k = 0$, the general solution is $y(x) = c_1 x + c_2$. These have at most one zero and so are nonoscillatory.

Next consider Airy's equation

$$y'' = xy, \tag{i}$$

which arises in various applications. Since (i) has a variable coefficient, we cannot study it by elementary methods; in particular, we cannot find the general solution as we did for the constant coefficient equation. We will instead first make a graphical study of the solutions of (i), and then study the solutions using the Sturm Comparison Theorem (see Chapter 9). Compute and plot several numerical solutions of (i) with different initial conditions $y(0) = c, y'(0) = d$.

To be specific, choose three different values for the pair c, d, and graph the corresponding solutions over the interval $[-10, 5]$. What do you observe? Do solutions have zeros on the negative x-axis? A few? Many? Do they get closer together as x becomes more negative? Do solutions have zeros on the positive x-axis? Where are the solutions oscillatory?

Information on the zeros of solutions of linear second order ODEs, and hence information on their oscillatory nature, can also be obtained from the Sturm Comparison Theorem. By comparing (i) with the equation $y'' - by = 0$ for $x \leq b < 0$, what do you learn about zeros on the negative x-axis and their spacing, and hence about the oscillatory nature of solutions of (i) for $x \leq 0$? By comparing (i) with $y'' = 0$ for $0 < x$, what do you learn about the oscillatory nature of solutions for $x > 0$? How many zeros could a solution have on the positive x-axis? Do the graphs you plotted above confirm these results?

17. In this problem, we study solutions of the parabolic cylinder equation

$$y'' + \left(n + \frac{1}{2} - \frac{x^2}{4} \right) y = 0,$$

which arises in the study of quantum-mechanical vibrations. Since the equation is unchanged if x is replaced by $-x$, any solution function will be symmetric with respect to the y-axis. Therefore, we focus our attention on $x \geq 0$.

(a) Find the corresponding first order equation for $z = \arctan(y/y')$, as described in the *Geometric Method* section of Chapter 9.

(b) For $n = 1$, plot the direction field for the z equation from $x = 0$ to $x = 10$. (Remember to use $-\pi/2 \leq z \leq \pi/2$.) Based on the plot, predict what the solutions y to the parabolic cylinder equation look like near $x = 0$, and for larger x. Is there a value of x around which you expect their behavior to change?

(c) Now numerically solve the parabolic cylinder equation for $n = 1$ with two sets of initial conditions $y(0) = 1, y'(0) = 0$ and $y(0) = 0, y'(0) = 1$. Plot the two solutions on the same graph. (It will probably help to use the **PlotRange** option.) Do the solutions behave as you expected?

(d) Repeat parts (b) and (c) for $n = 5$. Point out any differences from the case $n = 1$.

(e) Repeat parts (b) and (c) for $n = 15$. Discuss how the solutions are changing as n increases.

(f) Now consider the three solutions (for $n = 1, 5$, and 15) with $y(0) = 0$. Point out similarities and differences. By drawing an analogy with Airy's equation, argue from the direction fields that, for any n, exactly one solution function decays for large x while all others grow. Do you have enough graphical evidence from the numerical plots to conclude that the solution corresponding to the initial data $y(0) = 0$, $y'(0) = 1$ is that function? Why or why not? (*Hint:* Look to previous discussions of stability of solutions as a guide.)

18. In this problem, we study solutions of Bessel's equation of order n,

$$y'' + \frac{1}{x}y' + \left(1 - \frac{n^2}{x^2}\right)y = 0,$$

for $n > 0$. Solutions of this equation, called Bessel functions of order n, are used in the study of vibrations and waves with circular symmetry. Since the equation is unchanged if x is replaced by $-x$, we focus our attention on $x \geq 0$.

(a) Find the corresponding first order equation for $z = \arctan(y/y')$, as described in the *Geometric Method* section of Chapter 9.

(b) For $n = 1$, plot the direction field for the z equation from $x = 0$ to $x = 20$. (Remember to use $-\pi/2 \leq z \leq \pi/2$.) Based on the plot, predict what the Bessel functions of order 1 look like near $x = 0$, and for larger x. Is there a value of x around which you expect their behavior to change?

(c) Now plot the Bessel functions **BesselJ[n, x]** and **BesselY[n, x]** for $n = 1$ on the same graph. (It will probably help to use the **PlotRange** option.) Do the solutions behave as you expected?

(d) Repeat parts (b) and (c) for $n = 5$. Point out any differences from the case $n = 1$.

(e) Repeat parts (b) and (c) for $n = 15$. Discuss how the solutions are changing as n increases.

Chapter 10

Series Solutions

A primary theme of this book is that numerical, geometric, and qualitative methods can be used to study solutions of differential equations, even when we cannot find an exact formula solution. Of course, exact solutions are extremely valuable, and there are many techniques for finding them. For instance, techniques for finding exact solutions of second order linear differential equations include:

- The exponential substitution, which leads to solutions of an arbitrary constant coefficient homogeneous equation;
- The method of *reduction of order*, which produces a second linearly independent solution of a homogeneous equation when one solution is already known;
- The method of *undetermined coefficients*, which solves special kinds of inhomogeneous equations with constant coefficients; and
- The method of *variation of parameters*, which yields the solution of a general inhomogeneous equation, given a fundamental set of solutions to the homogeneous equation.

Each of these techniques reduces the search for a formula solution to a simpler problem in algebra or calculus: finding the root of a polynomial, computing an antiderivative, or solving a pair of simultaneous linear equations.

The process of finding formulas for exact solutions of equations, either by hand or by computer, is called *symbolic computation*. The **DSolve** command incorporates some of the techniques listed above. It enables us to find exact solutions more rapidly and more reliably than we could by hand.

There are many differential equations that do not yield to the techniques listed above. By using more advanced ideas from calculus, however, we can find exact solutions for a wider class of differential equations. In this chapter and the next, we discuss two such calculus-based techniques for finding exact solutions of differential equations: *Series Solutions* and *Laplace Transforms*. The method of Series Solutions enables us to solve differential equations with variable coefficients. The theory of Laplace Transforms (discussed in the next chapter) enables us to solve constant coefficient linear differential equations with discontinuous inhomogeneous terms. These are sophisticated techniques, involving improper integrals, complex variables, and power series.

Both techniques are computationally intensive. For each technique, we describe a short *Mathematica* program that automates the process of computing a symbolic solution to a differential equation or to an initial value problem. The methods discussed in these chapters are different from **DSolve**, because they involve a *sequence* of *Mathematica* commands. As with numerical and graphical methods, computationally intensive symbolic methods become more tractable when applied with the aid of a mathematical software system.

Series Solutions

Consider the second order homogeneous linear differential equation with variable coefficients

$$a(x)y''(x) + b(x)y'(x) + c(x)y(x) = 0. \tag{1}$$

Suppose the coefficient functions a, b, and c are *analytic* (i.e., have convergent power series representations) in a neighborhood of a point $x = x_0$. For simplicity, we shall assume that $x_0 = 0$, but everything we say remains valid for any point of analyticity. We begin by dividing by $a(x)$ to normalize the equation. This causes no problems as long as $a(0) \neq 0$. In this case, we refer to the origin as an *ordinary point* for the equation. Changing notation, we write

$$y''(x) + p(x)y'(x) + q(x)y(x) = 0. \tag{2}$$

Equations (1) and (2) have the same solutions. We will search for a power series solution of the form

$$y(x) = \sum_{n=0}^{\infty} a_n x^n. \tag{3}$$

If we can find such a solution, then $y(0) = a_0$ and $y'(0) = a_1$. Thus, if we are given initial values, then we can find the first two coefficients of the power series solution.

By substituting the infinite series (3) into equation (2), expanding p and q as power series, and combining terms of the same degree, we obtain a *recursion relation* for the coefficients a_n. In other words, we obtain algebraic equations for a_n that involve $a_0, a_1, \ldots, a_{n-1}$ and known coefficients of the power series of p and q. Sometimes, we can solve the recursion relation in *closed form* by finding an algebraic formula for a_n. This formula would involve known functions of n (such as $n!$ or powers of n) and a_0 and a_1, but no other coefficients of lower degree. If no initial data are given, then a solution to the recursion relation produces a general solution of the differential equation with arbitrary constants a_0 and a_1.

Solving a recursion relation in closed form is like finding an exact formula solution for a differential equation. In many cases, we cannot solve the recursion relation. Nevertheless, we can still use the recursion relation to compute as many coefficients as we wish. As long as we stay close to the origin, the leading terms should give a good approximation to the full power series solution. One must be careful, however, because the power series solution will be valid only inside its radius of convergence. The radius of convergence will be at least as large as the distance from the origin to the nearest singularity of p or q. When the recursion relation is solvable in closed form, we can use a standard test from calculus to compute the radius of convergence precisely. In general, we can only make an educated guess.

We can use *Mathematica* to automate the process of finding power series solutions. In fact, *Mathematica* can solve some recursion relations in closed form. You can explore that possibility by loading the package **DiscreteMath`RSolve`** and experimenting with **RSolve**. Here, however, we shall describe a straightforward technique for computing the leading coefficients explicitly.

EXAMPLE. Consider the initial value problem

$$y'' - xy' - y = 0, \qquad y(0) = 2, \qquad y'(0) = 1$$

from Problem 15 in Section 5.2 of Boyce & DiPrima. We will find the first five terms (*i.e.*, terms up to order four) in the power series expansion of the solution of this initial value problem.

The **Series** command generates a power series for an analytic function in a neighborhood of a point to any number of terms. For example, the command

```
Series[Exp[x], {x, 0, 4}]
```

gives the first five terms of the Taylor series for e^x, namely

$$1 + x + \frac{x^2}{2} + \frac{x^3}{6} + \frac{x^4}{24} + O(x)^5.$$

The symbol $O(x)^5$ means terms of degree at least 5. We can also produce a formal series. For example,

```
formalseries = Series[y[x], {x, 0, 4}]
```

yields

$$y(0) + y'(0)\,x + \frac{y''(0)\,x^2}{2} + \frac{y^{(3)}(0)\,x^3}{6} + \frac{y^{(4)}(0)\,x^4}{24} + O(x)^5.$$

We can even apply the **Series** command to the left-hand side of a differential equation, as follows:

```
odeseries = Series[y''[x] - x*y'[x] - y[x], {x, 0, 4}]
```

The result is

$$(-y(0) + y''(0)) + \left(-2\,y'(0) + y^{(3)}(0)\right)x + \left(\frac{-3\,y''(0)}{2} + \frac{y^{(4)}(0)}{2}\right)x^2 +$$

$$\left(\frac{-2\,y^{(3)}(0)}{3} + \frac{y^{(5)}(0)}{6}\right)x^3 + \left(\frac{-5\,y^{(4)}(0)}{24} + \frac{y^{(6)}(0)}{24}\right)x^4 + O(x)^5.$$

To produce a series solution to a differential equation, we first set the series expansion of the left-hand side of the differential equation to zero, incorporate the initial conditions, and apply the **Solve** command. The output will be a list of replacement rules containing the coefficients of the series solution

```
coeffs = Solve[{odeseries == 0, y[0] == 2, y'[0] == 1}]
```

We obtain

$$\{\{y(0) \to 2, y'(0) \to 1, y''(0) \to 2, y^{(3)}(0) \to 2, y^{(4)}(0) \to 6,$$
$$y^{(5)}(0) \to 8, y^{(6)}(0) \to 30\}\}$$

Now we use these replacement rules to substitute for the coefficients in a formal power series:

```
seriessol = Series[y[x], {x, 0, 5}] /. First[coeffs]
```

The result is

$$2 + x + x^2 + \frac{x^3}{3} + \frac{x^4}{4} + \mathrm{O}(x)^5.$$

We can use **Normal** to convert the series solution to a polynomial by removing the $\mathrm{O}(x)^5$.

```
answer = Normal[seriessol]
```

This produces the Taylor polynomial approximation of degree 4 to the solution of the initial value problem:

$$2 + x + x^2 + \frac{x^3}{3} + \frac{x^4}{4}.$$

The **DSolve** command will find exact solutions to many homogenous second order equations with polynomial coefficients. Often, these solutions involve special functions like **BesselJ** and **BesselY**. Equations with coefficients that are not polynomials generally require series methods.

Singular Points

In the discussion above, we considered equations of the form

$$a(x)y'' + b(x)y' + c(x)y = 0,$$

where $a(0) \neq 0$. We transformed the equation into the form

$$y'' + p(x)y' + q(x)y = 0$$

by dividing by $a(x)$. If $a(0) = 0$, then dividing by $a(x)$ may result in one or both of $p(x)$ and $q(x)$ being singular at $x = 0$. In this case, we say that the equation has a *singularity* at $x = 0$.

A prototype is the Euler equation:

$$x^2y'' + bxy' + cy = 0,$$

where b and c are constants. Dividing by x^2 yields the equation

$$y'' + \frac{b}{x}y' + \frac{c}{x^2}y = 0,$$

so $p(x) = b/x$ and $q(x) = c/x^2$. Both $p(x)$ and $q(x)$ are singular at $x = 0$. Such isolated singularities, where the singularity of p is no worse than $1/x$ and the singularity of q is no worse than $1/x^2$, are called *regular singular points*. More precisely, we say that a function $f(x)$ has a *pole of order n* at x_0 if $\lim_{x \to x_0} f(x) = \infty$, and n is the smallest integer such that $\lim_{x \to x_0} (x - x_0)^n f(x)$ is finite. A homogeneous linear differential equation $y^{(n)} + p_1(x)y^{(n-1)} + \ldots + p_n(x)y = 0$ is said to have a regular singular point at x_0 if it is singular at x_0, and p_k has a pole of order at most k at x_0.

Here is a summary of the general approach to solving second order linear equations with regular singular points. We suppose, for simplicity, that the singular point is $x = 0$, and we look for solutions on the interval $(0, \infty)$. Let $p_0 = \lim_{x \to 0} xp(x)$ and $q_0 = \lim_{x \to 0} x^2 q(x)$. Let r_1 and r_2 be the roots of the *indicial equation* $r(r-1) + p_0 r + q_0 = 0$. We look for a solution of the differential equation of the form

$$y_1(x) = x^{r_1} u(x),$$

where $u(0) = 1$. If $r_1 - r_2$ is not an integer, then we look for a second solution of the form

$$y_2(x) = x^{r_2} v(x),$$

where $v(0) = 1$. If $r_1 = r_2$, then we look for a second solution of the form

$$y_2(x) = y_1(x) \ln(x) + x^{r_1} v(x),$$

where $v(0) = 0$. If $r_1 - r_2$ is a positive integer, then we look for a solution of the form

$$y_2(x) = a y_1(x) \ln(x) + x^{r_1} v(x),$$

where a is a constant and $v(0) = 1$. The functions $u(x)$ and $v(x)$ will be analytic at 0 and therefore will have power series expansions at $x = 0$. These solutions are called *Frobenius series* solutions.

EXAMPLE. Consider the differential equation:

$$2xy'' + y' + xy = 0.$$

The origin is a regular singular point, and the indicial equation has roots $r = 0$ and $r = 1/2$. Hence, the differential equation should have a Frobenius series solution of the form

$$x^{1/2} \sum_{n=0}^{\infty} a_n x^n.$$

We will adapt the method above (developed for ordinary points) to compute the leading terms of the Frobenius series. We define $y(x)$ to be $\sqrt{x}\, u(x)$ and expand both y and the differential equation. Here is the sequence of commands:

```
y[x_] = Sqrt[x]*u[x]
odeseries = Series[2x*y''[x] + y'[x] + x*y[x], {x, 0, 5}]
coeffs = Solve[{odeseries == 0, u[0] == 1}]
seriessol = Series[y[x], {x, 0, 5}] /. First[coeffs]
answer = Normal[seriessol]
```

The final result is

$$\sqrt{x} - \frac{x^{5/2}}{10} + \frac{x^{9/2}}{360}.$$

EXERCISE. Compute two more terms in each of the examples presented in this chapter.

It is important to note that for equations with regular singular points, the Frobenius series tells us how fast the solution blows up at the singularity, *e.g.*, the solution blows up like $1/x$, $x^{-1/2}$, *etc.* This information is not easily gleaned from a numerical solution. For equations with an irregular singular point, we cannot expect a Frobenius series solution to be valid, and other techniques (numerical or qualitative) must be used. Some of those techniques are addressed in Problem Set E.

In this chapter, we have focused on linear homogeneous second order differential equations. The method of series solutions can also be used for inhomogeneous equations, higher order equations, and nonlinear equations. In fact, as we've noted before, most differential equations cannot be solved in terms of elementary functions. For many of these equations, a series solution is the only analytic solution available.

Chapter 11

Laplace Transforms

A *transform* is a mathematical operation that changes a given function into a new function. Transforms are often used in mathematics to change a difficult problem into a more tractable one. In this chapter, we introduce the *Laplace Transform*, which is particularly useful for solving linear differential equations with constant coefficients and discontinuous inhomogeneous terms. The key feature of the Laplace Transform is that (roughly speaking) it changes the operation of differentiation into the operation of multiplication. Thus the Laplace Transform changes a differential equation into an algebraic equation. To solve a linear differential equation with constant coefficients, you apply the Laplace Transform to change the differential equation into an algebraic equation, solve the algebraic equation, and then apply the Inverse Laplace Transform to transform the solution of the algebraic equation back into a solution of the differential equation.

The Laplace Transform of a function f is a new function, denoted by F or by $\mathcal{L}(f)$, and defined as follows:

$$F(s) = \mathcal{L}(f)(s) = \int_0^\infty f(t)e^{-st}\,dt.$$

This transform is called an *integral transform* because it is obtained by integrating the function f against another function e^{-st}, called the *kernel* of the transform. The integral in question is an improper integral, so we have to make sure that it converges. Note that while the argument s of the function F appears in the integrand, the integration is with respect to the variable t. Note also that the integral is over the domain $[0, \infty)$, so we need only assume that the function f is defined for $t \geq 0$.

To get a feel for the Laplace Transform, we compute the Laplace Transform of the function e^{at}.

$$\mathcal{L}(e^{at})(s) = \int_0^\infty e^{at}e^{-st}\,dt = \lim_{c \to \infty} \int_0^c e^{at}e^{-st}\,dt$$

$$= \lim_{c \to \infty} \int_0^c e^{(a-s)t}\,dt = \lim_{c \to \infty} \frac{e^{(a-s)t}}{(a-s)}\bigg|_0^c =$$

$$= \lim_{c \to \infty} \left(\frac{e^{(a-s)c}}{(a-s)} - \frac{1}{a-s} \right)$$

$$= \begin{cases} 1/(s-a), & \text{if } s > a \\ +\infty, & \text{if } s \le a. \end{cases}$$

Note that the Laplace Transform of e^{at} is defined for $s > a$. In fact, a straightforward argument shows that if f is any piecewise continuous function on $[0, \infty)$ with the property that $|f(t)| \le K e^{at}$, for some constant $K > 0$, then the improper integral defining the Laplace Transform converges for $s > a$, and therefore $\mathcal{L}(f)(s)$ is defined for $s > a$. We say that a function is *piecewise continuous* if it only has a discrete set of jump discontinuities. If f satisfies an inequality of the form $|f(t)| \le K e^{at}$, we say that f is of *exponential order*. Most functions one encounters in practice are of exponential order. From now on, we only consider piecewise continuous functions of exponential order. In particular, if f is a bounded function, then it satisfies the inequality $|f(t)| \le K = K e^{0t}$ for some $K > 0$, so $\mathcal{L}(f)(s)$ is defined for all $s > 0$. More generally, any function whose growth is of exponential order has a Laplace Transform which is defined for sufficiently large s.

EXERCISE. Compute the Laplace Transform of the functions $f(t) = 1$ and $\cos t$. (For $\cos t$, you must integrate by parts twice.)

We asserted that the Laplace Transform changes differentiation into multiplication. This is a consequence of the integration by parts formula:

$$\begin{aligned}
\mathcal{L}(f')(s) &= \int_0^\infty f'(t) e^{-st}\, dt = \lim_{c \to \infty} \int_0^c f'(t) e^{-st}\, dt \\
&= \lim_{c \to \infty} \left[f(t) e^{-st} \Big|_0^c - \int_0^c f(t)(-se^{-st})\, dt \right] \\
&= -f(0) + s \lim_{c \to \infty} \int_0^c f(t) e^{-st}\, dt \\
&= s\mathcal{L}(f)(s) - f(0).
\end{aligned} \tag{1}$$

To be precise, we must assume that f is differentiable and f' is piecewise continuous for this formula to hold. We can summarize (1) as follows: If the Laplace Transform of f is $F(s)$, then the Laplace Transform of f' is $sF(s) - f(0)$. In other words, the Laplace Transform changes the operation of differentiation into the operation of multiplication (by the independent variable) plus a translation (by $-f(0)$). The formula in (1) has a straightforward generalization to higher derivatives. If the Laplace Transform of f is F, $f^{(k)}$ is continuous for $k = 0 \ldots n-1$, and $f^{(n)}$ is piecewise continuous, then

$$\mathcal{L}(f^{(n)})(s) = s^n F(s) - s^{n-1} f(0) - s^{n-2} f'(0) - \cdots - f^{(n-1)}(0).$$

The Laplace Transform has many other important properties, of which we mention two here. First, the Laplace Transform is linear; *i.e.*, $\mathcal{L}(af+bg) = a\mathcal{L}(f)+b\mathcal{L}(g)$.

Linearity of the Laplace Transform follows easily from linearity of integration. Second, the Laplace Transform is invertible; the inverse is called the Inverse Laplace Transform and is denoted by $\mathcal{L}^{-1}$. The Inverse Laplace Transform has the property that $\mathcal{L}^{-1}(\mathcal{L}(f)) = f$; *i.e.*, the Inverse Laplace Transform of the Laplace Transform of a function is the function itself. The Inverse Laplace Transform is also an integral transform, but it involves contour integrals in the complex plane, so we do not give the definition here.

Solving Differential Equations with Laplace Transforms

Let's see what happens when we apply the Laplace Transform to a second order linear differential equation with constant coefficients. Consider the initial value problem

$$ay''(t) + by'(t) + cy(t) = f(t), \qquad y(0) = y_0, \qquad y'(0) = y_0'. \tag{2}$$

When we apply the Laplace Transform to this equation, we get the algebraic equation

$$a(s^2 Y(s) - s y_0 - y_0') + b(s Y(s) - y_0) + c Y(s) = F(s),$$

where $Y(s)$ is the Laplace Transform of $y(t)$ and $F(s)$ is the Laplace Transform of $f(t)$. We solve this algebraic equation for $Y(s)$ to get

$$Y(s) = \frac{F(s) + a s y_0 + a y_0' + b y_0}{a s^2 + b s + c},$$

and then compute the Inverse Laplace Transform of the right-hand side to get an expression for $y(t)$.

Traditionally, one used tables to look up Laplace Transforms and Inverse Laplace Transforms. You can also use *Mathematica* to compute Laplace Transforms. First, you must load the Laplace Transform package by typing

```
<<Calculus`LaplaceTransform`
```

The Laplace Transform of a function $y(t)$ is then computed by

```
LaplaceTransform[y[t], t, s]
```

and the inverse Laplace Transform of a function $Y(s)$ is computed by

```
InverseLaplaceTransform[Y[s], s, t]
```

For example, **LaplaceTransform[Cos[t], t, s]** yields $s/(1 + s^2)$.

We can use these commands to implement the Laplace Transform method for solving differential equations. Consider the initial value problem $y'' + y = \sin 2t$, $y(0) = 1$, $y'(0) = 0$. First, we define the equation we want to solve:

```
eqn = y''[t] + y[t] == Sin[2t]
```

Then we compute the Laplace Transform of this equation:

```
LaplaceTransform[eqn, t, s]
```

which produces

$$\mathrm{LaplaceTransform}[y[t], t, s] + s^2\, \mathrm{LaplaceTransform}[y[t], t, s] -$$
$$s\, y[0] - y'[0] == \frac{2}{4 + s^2}$$

To simplify matters, we can replace **LaplaceTransform[y[t], t, s]** with **Y[s]**, as is done in most textbooks.

```
lteqn = LaplaceTransform[eqn, t, s] /.
  LaplaceTransform[y[t], t, s] -> Y[s]
```

This yields

$$-s\, y[0] + Y[s] + s^2\, Y[s] - y'[0] == \frac{2}{4 + s^2}$$

Now we can solve this equation, together with the initial conditions, for $Y[s]$:

```
ltsol = Solve[{lteqn, y[0] == 1, y'[0] == 0}, Y[s]]
```

which yields

$$\{\{Y[s] \rightarrow \frac{s + \frac{2}{4+s^2}}{1 + s^2}\}\}$$

Finally, we apply the Inverse Laplace Transform to the right-hand side of the preceding expression:

```
InverseLaplaceTransform[Y[s] /. First[ltsol], s, t]
```

which yields the solution of the initial value problem:

$$\cos(t) + 2\left(\frac{\sin(t)}{3} - \frac{\sin(2\,t)}{6}\right).$$

We can bundle this procedure into a simple *Mathematica* command as follows:

```
<<Calculus`LaplaceTransform`
LTSolve[eqn_, y0_, yp0_] := Module[{lteqn, ltsol, Y, s},
  lteqn = LaplaceTransform[eqn, t, s] /.
    LaplaceTransform[y[t], t, s] -> Y[s];
  ltsol = Solve[{lteqn, y0, yp0}, Y[s]];
  InverseLaplaceTransform[Y[s] /. First[ltsol], s, t]]
```

We can apply this command to the initial value problem above by entering

```
LTSolve[y''[t] + y[t] == Sin[2t], y[0] == 0, y'[0] == 1]
```

NOTE. The **Module** command in the definition of **LTSolve** is used to specify that the listed variables (**lteqn**, **ltsol**, **s**, and **Y**) are *local* to the function definition. This prevents conflicts that might arise if some of these variables have been previously defined. Note that **t** and **y** are not local to the function definition and *should not* be included in the list of local variables.

When using the **LTSolve** command, it is essential that you use **t** as the independent variable and **y** as the dependent variable in the differential equation. The **LTSolve** command works for second order equations; for higher order equations you must modify the definition to include additional initial conditions.

EXAMPLE. Consider the initial value problem:

$$y''(t) - 2y(t) = \sin 2t, \quad y(0) = 1, \quad y'(0) = 2.$$

Typing

```
eqn2 = y''[t] - 2*y[t] == Sin[2t]
LTSolve[eqn2, y[0] == 1, y'[0] == 2]
```

produces the solution

$$\cosh(\sqrt{2}\,t) + \sqrt{2}\,\sinh(\sqrt{2}\,t) + 2\left(\frac{-\sin(2\,t)}{12} + \frac{\sinh(\sqrt{2}\,t)}{6\sqrt{2}}\right).$$

Discontinuous Functions

The Laplace Transform is especially useful for solving differential equations that involve piecewise continuous functions. The basic building block for piecewise continuous functions is the *unit step function* $u(t)$, defined by

$$u(t) = \begin{cases} 0, & \text{if } t < 0, \\ 1, & \text{if } t \geq 0. \end{cases}$$

In honor of the physicist and engineer Oliver Heaviside (1850-1925), who developed the Laplace Transform method to solve problems in electrical engineering, this function is sometimes called the Heaviside function. In *Mathematica*, this function is called **UnitStep**. It is in the **Calculus`LaplaceTransform`** package.

The unit step function is best thought of as a switch, which is off until time 0, and then on starting at time 0. The function $1 - u(t)$ is also a switch, on until time 0, and then off. To make a switch that comes on at time c, we simply translate $u(t)$ by c; thus $u(t - c)$ is a switch that comes on at time c. Similarly, $1 - u(t - c)$ is a switch that goes off at time c. The unit step function translated by c is often written $u_c(t) = u(t-c)$. Note that $1 - u(t-c) = u(c-t)$, or equivalently, $1 - u_c(t) = u_{-c}(-t)$.

The unit step function can be used to build piecewise continuous functions by switching pieces of the function on and off at appropriate times. To turn a function on at time c, multiply it by $u_c(t)$, and to turn it off at time d, multiply it by $(1 - u_d(t))$. Consider, for example, the function

$$f(t) = \begin{cases} 0, & t < 0, \\ 1, & 0 \leq t < 1, \\ t^2, & 1 \leq t < 3, \\ \sin 2t, & t \geq 3. \end{cases}$$

We can write this as

$$f(t) = u_0(t)(1 - u_1(t)) + u_1(t)(1 - u_3(t))\,t^2 + u_3(t)\sin 2t.$$

In *Mathematica*, we would enter this as

```
f[t_] = UnitStep[t]*UnitStep[1 - t] +
  UnitStep[t - 1]*UnitStep[3 - t]*t^2 +
  UnitStep[t - 3]*Sin[2t];
```

Note that we used the identity `UnitStep[c - t] = 1 - UnitStep[t - c]`. Here is a plot of f.

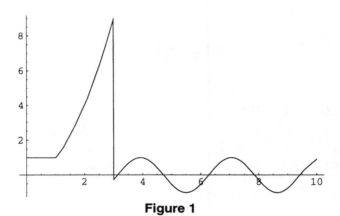

Figure 1

In this plot, you can clearly see the three different "pieces" of the function. (The vertical segment in the graph is an artifact of *Mathematica*'s plotting routine, and is not part of the function.)

EXERCISE.   Write down an expression for the following function in terms of the `UnitStep` function, and then plot it:

$$f(t) = \begin{cases} 0, & t < 1, \\ \cos t, & 1 \le t < 2, \\ \sin t, & t \ge 2. \end{cases}$$

Another important discontinuous function is the *Dirac delta function*, usually denoted $\delta(t)$ or $\delta_0(t)$, and sometimes called the *unit impulse function*. To understand this function, recall that the total impulse imparted by a varying force $F(t)$ is the integral of $F(t)$. If we consider a force of total impulse 1 and make this force act over a smaller and smaller time interval around 0, then in the limit we obtain the Dirac delta function. Thus, the delta function represents an idealized force of total impulse 1 concentrated at the instant $t = 0$. The delta function belongs to a class of mathematical objects called *generalized functions*. (It is not a true function since its value at 0 would be infinite.) The magnitude and timing of the delta function can be changed by multiplying the delta function by a constant and translating it. Thus, the function $10\delta(t - 8)$ represents a force of total impulse 10 concentrated at the instant $t = 8$. The function $\delta(t - c)$ is sometimes denoted $\delta_c(t)$. In *Mathematica*, the Dirac delta function is called **DiracDelta**. It is in the **Calculus`LaplaceTransform`** package.

EXERCISE. The fact that **DiracDelta** is not a true function causes unusual behavior in *Mathematica*. Plot the Dirac function on the interval $[-1, 1]$ (you won't see anything). Evaluate the function at 0 and 1. Integrate the function over the interval $[-1, 1]$.

Both the unit step function and the Dirac delta function have Laplace Transforms. For example,

$$\mathcal{L}(u_c)(s) = \int_0^\infty u_c(t)e^{-st}\, dt = \int_c^\infty e^{-st}\, dt$$

$$= \frac{e^{-cs}}{s}, \quad \text{for } s > 0.$$

To compute the Laplace Transform of $\delta(t)$, we must write $\delta(t)$ as a limit of *bona fide* functions. The easiest way to do this is to use step functions. We can write a function of total integral 1, concentrated on the interval $[-\epsilon, \epsilon]$, as

$$\frac{1}{2\epsilon}(u_{-\epsilon}(t) - u_\epsilon(t)).$$

Thus

$$\delta(t) = \lim_{\epsilon \to 0+} \frac{1}{2\epsilon}(u_{-\epsilon}(t) - u_\epsilon(t)),$$

and

$$\mathcal{L}(\delta_c)(s) = \lim_{\epsilon \to 0+} \frac{1}{2\epsilon}(\mathcal{L}(u_{c-\epsilon})(s) - \mathcal{L}(u_{c+\epsilon})(s))$$

$$= \lim_{\epsilon \to 0+} \frac{1}{2\epsilon}(\frac{e^{(\epsilon-c)s}}{s} - \frac{e^{-(\epsilon+c)s}}{s})$$

$$= \lim_{\epsilon \to 0+} (e^{-cs})\frac{e^{\epsilon s} - e^{-\epsilon s}}{2\epsilon s}$$

$$= (e^{-cs}) \lim_{\epsilon \to 0+} \frac{s(e^{\epsilon s} + e^{-\epsilon s})}{2s}$$

$$= (e^{-cs}) \lim_{\epsilon \to 0+} \cosh(\epsilon s)$$

$$= e^{-cs}.$$

(We used L'Hôpital's rule in passing from the third to the fourth line.) Thus $\mathcal{L}(\delta_c)(s) = e^{-cs}$. In particular, setting $c = 0$, we get that $\mathcal{L}(\delta)(s) = 1$.

Differential Equations with Discontinuous Forcing Functions

Consider an inhomogeneous, second order linear equation with constant coefficients:

$$ay''(t) + by'(t) + cy(t) = g(t).$$

The function $g(t)$ is called the forcing function of the differential equation, because in many physical models $g(t)$ corresponds to the influence of an external force. If $g(t)$ is piecewise continuous or involves the delta function, then we can solve the equation by the method of Laplace Transforms.

EXAMPLE. Consider the initial value problem:

$$y''(t) + 3y'(t) + y(t) = g(t), \quad y(0) = 1, \quad y'(0) = 1,$$

where

$$g(t) = \begin{cases} 0, & t < 0, \\ 1, & 0 \le t < 1, \\ -1, & 1 \le t < 2, \\ 0, & t \ge 2. \end{cases}$$

We can solve this equation in *Mathematica* with the following sequence of commands.

```
g[t_] = UnitStep[t] - 2UnitStep[t - 1] + UnitStep[t - 2]
eqn3 = y''[t] + 3*y'[t] + y[t] == g[t]
sol3[t_] = LTSolve[eqn3, y[0] == 1, y'[0] == 1]
```

The solution is too complicated to reproduce here (but you should compute it using *Mathematica*). Here is a graph of the solution together with the forcing function, produced with the command **Plot[{sol3[t], g[t]}, {t, 0, 5}]**.

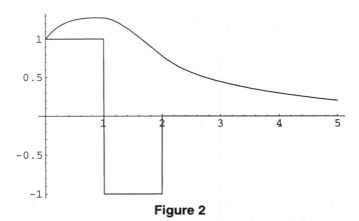

Figure 2

EXAMPLE. We can also use *Mathematica* to solve nonhomogeneous equations involving the delta function. Consider the initial value problem

$$y''(t) + y'(t) + y(t) = \delta_1(t), \quad y(0) = 0, y'(0) = 0.$$

We can solve this equation with the following sequence of commands.

```
eqn4 = y''[t] + y'[t] + y[t] == DiracDelta[t - 1]
sol4[t_] = LTSolve[eqn4, y[0] == 0, y'[0] == 0]
```

The solution is

$$\left(\frac{-i}{\sqrt{3}\,e^{\frac{(1-i\sqrt{3})\,(-1+t)}{2}}} + \frac{i}{\sqrt{3}\,e^{\frac{(1+i\sqrt{3})\,(-1+t)}{2}}} \right) \text{UnitStep}[-1 + t].$$

Here is a graph of the solution.

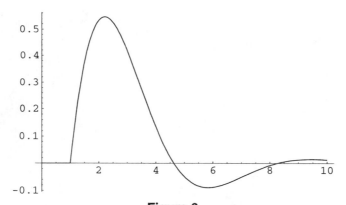

Figure 3

In this graph, you can clearly see the effect of the unit impulse at time $t = 1$. Until $t = 1$, the solution is 0. At time $t = 1$, the unit impulse puts the solution into a nonzero state. After $t = 1$, the impulse function has no further effect. Since the roots of the characteristic polynomial of the homogeneous equation have negative real part, the solution decays exponentially to 0.

Problem Set E

Series Solutions and Laplace Transforms

The solutions to Problems 1 and 16 appear in the *Sample Notebook Solutions*.

1. Consider Airy's equation $y'' - xy = 0$.

    (a) Compute the terms of degree 10 or less in the Taylor series expansion of the solution $y(x)$ to Airy's equation $y'' - xy = 0$ that satisfies $y(0) = 1$, $y'(0) = 0$.

    (b) On the intervals $(-10, 0)$ and $(0, 5)$, graph the Taylor polynomial you obtained, together with the exact solution of the initial value problem. (You can compute the exact solution with **DSolve**. The solution will be given in terms of the built-in functions **AiryAi** and **AiryBi**.) Where is the approximation accurate?

    (c) Airy's equation can be written as $y'' = xy$. If $x > 1$, then $y'' = xy > y$, so the solutions of Airy's equation grow at least as quickly as the solutions of $y'' = y$. Analyze the latter equation, and predict how well Taylor polynomials will approximate exact solutions of Airy's equation. Relate your analysis to the graph in part (b).

2. Compute the terms of degree 10 or less in the Taylor series expansion of the solution to the following differential equation (*cf.* Problem 23, Section 5.2 of Boyce & DiPrima):

$$y'' - xy' - y = 0, \quad y(0) = 1, \quad y'(0) = 0.$$

    (a) Graph the Taylor polynomial of degree 10.

    (b) What is the long-term behavior of the Taylor polynomial (as $x \to \pm\infty$)?

    (c) Is the Taylor polynomial an even function? odd? neither?

    (d) Explain purely from the IVP itself why each of your conclusions holds for the actual solution. (*Hints*: In (b), you will have to do a little qualitative analysis—approximate x by a constant; in (c), consider $y_1(x) = y(-x)$.)

3. Compute the terms of degree 10 or less in the Taylor series expansion of the solution of the following differential equation (*cf.* Problem 18, Section 5.2 of

Boyce & DiPrima):

$$(1 - x)y'' + xy' - y = 0, \quad y(0) = -3, \quad y'(0) = 2.$$

The theory suggests that the true solution may have a singularity at $x = 1$. Graph the Taylor polynomial you obtained. Do you see any signs of the singularity? Compute the exact solution using **DSolve**. Does the solution have a singularity?

4. Use **DSolve** to solve the following Euler equations (*cf.* Section 5.5 of Boyce & DiPrima, Problems 3, 6, 8, 13):

   (a) $x^2 y'' - 3xy' + 4y = 0$

   (b) $(x - 1)^2 y'' + 8(x - 1)y' + 12y = 0$

   (c) $2x^2 y'' - 4xy' + 6y = 0$

   (d) $2x^2 y'' + xy' - 3y = 0, \quad y(1) = 1, \quad y'(1) = 4.$

5. Use the program for a Frobenius series expansion in Chapter 10 to compute the terms of degree 10 or less in the Frobenius series expansion of the solution to the following problems in Section 5.6 of Boyce & DiPrima (Problems 2, 5, 8):

   (a) $x^2 y'' + xy' + (x^2 - \frac{1}{9})y = 0, \quad r = -\frac{1}{3}$

   (b) $3x^2 y'' + 2xy' + x^2 y = 0, \quad r = \frac{1}{3}$

   (c) $2x^2 y'' + 3xy' + (2x^2 - 1)y = 0, \quad r = \frac{1}{2}.$

6. This problem concerns Bessel's equation, which arises in many physical problems with circular symmetry, such as the study of water waves in a circular pond, vibrations of a circular drum, or the diffraction of light by the circular aperture of a telescope. The Bessel function $J_n(x)$ (where the parameter n is an arbitrary integer) is defined to be the coefficient of t^n in the power series expansion of the function

$$\exp\left(\frac{1}{2}x\left(t - \frac{1}{t}\right)\right).$$

(By thinking about what happens to the power series when t and $1/t$ are switched, you can see that $J_{-n}(x) = J_n(x)$ for n even and $J_{-n}(x) = -J_n(x)$ for n odd.) $J_n(x)$ is also a solution to *Bessel's equation of order n*,

$$x^2 y'' + xy' + (x^2 - n^2)y = 0.$$

The Taylor series expansion for J_0 is (*cf.* Section 5.8 of Boyce & DiPrima):

$$J_0(x) = \sum_{j=0}^{\infty}(-1)^j \frac{1}{2^{2j}} \frac{x^{2j}}{(j!)^2}.$$

   (a) *Mathematica* has a built-in function for $J_n(x)$, called **BesselJ[n, x]**. Use it to plot $J_0(x)$ for $0 \le x \le 10$. (You don't need to graph the function for negative x since J_0 is an even function.) On the same axes, draw the graphs of the 4th order, 10th order, and 20th order Taylor polynomials of J_0. (Use the **Series** command.) How well do the Taylor polynomials approximate the function?

(b) An interesting equality satisfied by the Bessel functions is

$$J'_n(x) = \frac{1}{2}\left(J_{n-1}(x) - J_{n+1}(x)\right).$$

In particular,

$$J'_0(x) = \frac{1}{2}\left(J_{-1}(x) - J_1(x)\right) = \frac{1}{2}\left(-J_1(x) - J_1(x)\right) = -J_1(x).$$

Thus, the maxima and minima of J_0 occur at the zeros of J_1. Using **FindRoot**, solve the equations $J_0(x) = 0$ and $J_1(x) = 0$ numerically to find the first five zeros and the first five relative extreme points of J_0, starting at $x = 0$. (You will need five different starting values—use your graph for guidance.) Compute the differences between successive zeros. Can you make a guess about the periods of the oscillations as $x \to \infty$? (If you did Problem 15 in Problem Set D, you know exactly how the period behaves as $x \to \infty$.)

(c) From the general theory, one knows that Bessel's equation of order n has another linearly independent solution $Y_n(x)$, with a logarithmic singularity as $x \to 0^+$. *Mathematica* also has a built-in function, **BesselY[n, x]**, for computing $Y_n(x)$. Graph the function $Y_0(x)$ on the interval $0 < x \leq 10$. Then compute $\lim_{x \to 0^+} Y_0(x)/\ln x$. Graph the function

$$Y_0(x) - c\ln x$$

(on the same interval), where c is the value of the limit you obtained, and observe that it is continuous in a neighborhood of $x = 0$. Thus, the "singular part" of $Y_0(x)$ behaves like $c\ln x$.

7. The first order homogeneous linear differential equation

$$x^2 y' + y = 0$$

has an irregular singular point at $x = 0$ (why?); however, it happens to be an equation that is easy to solve exactly. Find the general solution of the equation for $x > 0$ and for $x < 0$. How do the solutions behave as $x \to 0^+$? as $x \to 0^-$? Use *Mathematica* to graph a solution for $0 < x < 2$ and a solution for $-2 < x < 0$.

Now find (for $x > 0$ and for $x < 0$) the general solution of the Euler equation

$$xy' + ry = 0,$$

which has a regular singular point at $x = 0$ if $r \neq 0$. How do the solutions behave as $x \to 0^+$? as $x \to 0^-$? Use *Mathematica* to graph solutions to the right and left of 0 for each of the cases $r = 0.5$, $r = 1$, $r = -0.5$, $r = -1$.

What difference do you notice between the behavior of solutions near a regular singular point and the behavior near an irregular singular point?

8. Consider the Legendre equation

$$(1 - x^2)y'' - 2xy' + \alpha(\alpha + 1)y = 0.$$

We assume that $|x| < 1$, because $x = \pm 1$ are regular singular points. We also assume that $\alpha > -1$.

(a) Find series approximations of the two solutions of the Legendre equation with initial conditions $y(0) = 1$, $y'(0) = 0$, and $y(0) = 0$, $y'(0) = 1$. Is this a fundamental set of solutions?

(b) If $\alpha = n$ is a nonnegative integer, the Legendre equation has a polynomial solution of order n. The *Legendre polynomial* $P_n(x)$ is defined to be the unique polynomial solution of the Legendre equation (with $\alpha = n$) such that $P_n(1) = 1$. Use *Mathematica* to compute the first six Legendre polynomials $P_0(x), \ldots, P_5(x)$. You can do this by using **DSolve** (without initial conditions) to solve the Legendre equation for $n = 0, \ldots, 5$; then by inspecting each solution, choose values for the undetermined constants to get polynomial solutions with $P_n(1) = 1$.

(c) Graph $P_0(x), \ldots, P_5(x)$ on the interval $[-1, 1]$.

(d) In the graph in part (c), you can see that as n increases, the Legendre polynomials are more and more oscillatory on the interval $[-1, 1]$. For x close to zero, the Legendre equation is approximately $y'' + n(n+1)y = 0$. By solving this equation, explain why solutions of the Legendre equation display increasing oscillation (on the interval $[-1, 1]$) as $n \to \infty$.

9. Nonlinear differential equations can sometimes be solved by series methods. Consider, for example, the equation

$$y' = y^2 + x, \qquad (i)$$

mentioned in Chapter 7 as an example of an equation that cannot be solved in terms of elementary functions. Consider this equation with the initial condition $y(0) = 0$, and let's look for a solution of the form

$$y(x) = \sum_{i=1}^{\infty} a_i x^i. \qquad (ii)$$

(We know there is no constant term in the Taylor series expansion of y at $x = 0$ because of the initial condition.) Substituting (ii) into (i) and expanding the right-hand side give a sequence of simultaneous quadratic equations for the coefficients a_i that become increasingly complicated as $i \to \infty$. However, we can solve inductively for the coefficients, by first equating the terms in x, then the terms in x^2, etc. *Mathematica* can assist us in this task.

(a) Explain why a_1 must be 0.

(b) Now use the scheme from Chapter 10 to compute a Taylor series expansion of the solution to (i). Make sure you go far enough to get the 11th degree monomial. What pattern do you see?

(c) Plot the 11th degree Taylor polynomial for y, together with a numerical solution of (i) obtained using **NDSolve**, on the interval $-1 < x < 1$. How close are the graphs? Now note that $y(x) > x^2/2$ for $0 < x < 1$, as you can see, for example, from the series. Thus, $y(1) > 0.5$. But for $x > 1$, the right-hand side of (i) is greater than $y(x)^2 + 1$, so the solution grows faster than the solution to the IVP $y' = y^2 + 1$, $y(1) = 0.5$, which is of the form $y(x) = \tan(x - C)$, where $C \approx 0.54$. Since the tangent function blows up when its argument reaches $\pi/2$, this analysis shows that the solution of (i) must become infinite somewhere between $x = 1$ and $x = 2.2$. Is there any way of seeing this from the series solution? Use **NDSolve** to locate the point P where the solution blows up. Graph the two approximate solutions (that is, the series solution out to degree 11 and the numerical solution on the interval $(-1, P)$). What do you observe?

10. Suppose $f(t)$ and $g(t)$ are functions with Laplace Transforms $\mathcal{L}(f) = F$ and $\mathcal{L}(g) = G$. Show that *Mathematica* knows all of the following identities. (For (b), you will have to use a fixed positive number rather than an arbitrary constant c.)

(a) $\mathcal{L}(af + bg) = a\mathcal{L}(f) + b\mathcal{L}(g)$, where a and b are real numbers.

(b) $\mathcal{L}(u_c(t)f(t - c)) = e^{-cs}F(s)$

(c) $\mathcal{L}(e^{ct}f(t)) = F(s - c)$

(d) $\mathcal{L}(f') = sF(s) - f(0)$

(e) $\mathcal{L}(\int_0^t f(t - u)g(u)\,du) = F(s)G(s)$.

The function $t \to \int_0^t f(t - u)g(u)\,du$ is called the *convolution* of f and g, and is written $f * g$. Thus, part (e) says that the Laplace Transform changes convolution into ordinary multiplication.

11. For each of the following functions, compute the Laplace Transform. Then plot the function and its Laplace Transform over the interval $[0, 10]$. You may want to restrict the vertical range for some of the graphs, and to use different plot styles for the function and its Laplace Transform.

(a) $\sin t$

(b) e^t

(c) $\dfrac{1}{t + 1}$

(d) $\cos t$

(e) $t \cos t$

(f) $u_1(t) \sin t$

On the basis of these graphs, what general conclusions can you draw about the Laplace Transform of a function? In particular, what can you say about the growth or decay of the Laplace Transform of a function? Can you justify your conclusion by looking at the integral formula for the Laplace Transform? (*Hint*: Look at the derivation of $\mathcal{L}(e^{at})$ on page 140.)

12. This problem illustrates how the choice of method can dramatically affect the time it takes the computer to solve a differential equation. It can also affect the form of the solution. Consider the IVP

$$y'' + y' + y = t^2 e^{-t} \cos t, \qquad y(0) = 1, \qquad y'(0) = 0.$$

This problem could be solved by the method of undetermined coefficients, variation of parameters, or Laplace Transforms.

(a) Try to solve the IVP using **DSolve**. If you have a reasonably fast computer with sufficient memory and are using *Mathematica* 3.0, then **DSolve** will eventually return a solution, though the form of the solution will be quite complicated. (If you really have some time to kill, you can try to **Simplify** the solution.)

(b) Now use the **LTSolve** function from Chapter 11 to solve the problem. Use **Simplify** to simplify the answer.

(c) Verify that the expression produced in part (b) is really a solution to the differential equation (don't forget **Simplify**). Plot the solution.

13. Consider the following IVP:

$$y'' + y' + \frac{5}{4}y = g(t), \quad y(0) = 0, \quad y'(0) = 0, \quad g(t) = \begin{cases} \sin t, & \text{if } 0 \le t < \pi \\ 0, & \text{if } t \ge \pi. \end{cases}$$

(a) Solve this equation using **DSolve**. Plot the solution on the interval $[0, 15]$.

(b) Check to see whether the function produced by **DSolve** is actually a solution by substituting it back into the differential equation. The expression will be quite complicated. Evaluate it numerically at $t = 0.5$, 1, and 3, to see if both sides of the equation are the same.

(c) Now solve the equation using the Laplace Transform method. Plot the solution on $[0, 15]$. Check to see whether this solution is correct by substituting into the original equation. As in part (b), you will have to check by evaluating both sides of the equation numerically at several points, because *Mathematica* cannot completely simplify the solution.

(d) Find the general solution of the associated homogeneous equation. What is the asymptotic behavior of those solutions as $t \to \infty$? Knowing that the forcing term of the inhomogeneous equation is zero after $t = \pi$, what can you say about the long-term behavior of the solution of the inhomogeneous equation?

14. Use the Laplace Transform method to solve the following IVPs. Then graph the solution together with the function on the right-hand side of the differential equation (the forcing function) on the interval $[0, 15]$.

(a) $y'' + 2y' + 2y = h(t)$, $y(0) = 0$, $y'(0) = 1$, $h(t) = \begin{cases} 1, & \text{if } \pi \le t < 2\pi \\ 0, & \text{otherwise} \end{cases}$

(b) $y'' + 3y' + 2y = u_2(t)$, $y(0) = 0$, $y'(0) = 1$

(c) $y'' + y' + \dfrac{5}{4}y = g(t)$, $y(0) = 0$, $y'(0) = 0$, $g(t) = \begin{cases} \sin t, & \text{if } 0 \le t < \pi \\ 0, & \text{if } t \ge \pi \end{cases}$

(d) $y'' + y = \delta(t - \pi)\cos t$, $y(0) = 0$, $y'(0) = 1$

(e) $y'' + 2y' + 2y = \cos t + \delta(t - \pi/2)$, $y(0) = 0$, $y'(0) = 0$.

15. Use the Laplace Transform method to solve the following IVPs. (You will have to modify the **LTSolve** function from Chapter 11 to use it on third and fourth order equations, by adding arguments **ypp0** and **yppp0** for the extra initial conditions.) Then plot the solution on an appropriate interval $(t > 0)$.

(a) $y'''(t) - y''(t) - y'(t) + y(t) = \delta(t - 1)$, $y(0) = y''(0) = 0$, $y'(0) = 1$

(b) $y^{(4)}(t) + 2y''(t) + y(t) = \sin t$, $y(0) = 1$, $y'(0) = y''(0) = y'''(0) = 0$

(c) $y^{(4)}(t) + 5y''(t) + 4y(t) = u_1(t)\sin(3t)$, $y(0) = y'(0) = y''(0) = y'''(0) = 0$

(d) $y'''(t) + y''(t) + y'(t) = \delta(t - 1)$, $y(0) = y''(0) = 0$, $y'(0) = 1$.

16. In this problem we investigate the effect of a periodic discontinuous forcing function (a square wave) on a second order linear equation with constant coefficients. Consider the initial value problem

$$y'' + y = h(t), \quad y(0) = 0, \, y'(0) = 1. \tag{i}$$

The associated homogeneous equation has natural period 2π; the general solution of the homogeneous equation is

$$y(t) = A\cos t + B\sin t.$$

Recall that the phenomenon of resonance occurs when the forcing function $h(t)$ is a linear combination of $\sin t$ and $\cos t$. Does resonance occur when the forcing function is periodic of period 2π but discontinuous?

(a) Using step functions, define a *Mathematica* function $h(t)$ on the interval $[0, 10\pi]$ whose value is $+1$ on $[0, \pi)$, -1 on $[\pi, 2\pi)$, $+1$ on $[2\pi, 3\pi)$, and so on. Plot the function on the interval $[0, 30]$. It should have the appearance of a square wave.

(b) Use the program for the Laplace Transform method from Chapter 11 to solve equation (i) with the function $h(t)$ defined in part (a). Plot the solution together with $h(t)$ on the interval $[0, 30]$. Do you see resonance? Compute and plot a numerical solution to confirm your answer.

(c) In part (a), we constructed a forcing function $h(t)$ with period 2π. The function $h(t/2)$ has period 4π. Repeat part (b) using the forcing function $h(t/2)$. Do you see resonance?

(d) Repeat part (b) using the forcing function $h(2t)$ (this time just plot from 0 to 15). Do you see resonance? What is the period of $h(2t)$?

(e) What can you conclude about the resonance effect for discontinuous forcing functions? Would you expect resonance to occur in equation (i) for *any* forcing function of period 2π? (*Hint:* The function $h(2t)$ has period π as well as period 2π.) Can you venture a guess about when resonance occurs for discontinuous periodic forcing functions? You might try some other periodic forcing functions to check your guess.

17. Do Problem 16 with a sawtooth wave instead of a square wave. More specifically, replace part (a) of Problem 16 with the new part (a) below, and then do parts (b)–(e) of Problem 16.

(a) Using step functions, define a *Mathematica* function $h(t)$ on the interval $[0, 10\pi]$, with $h(t) = t$ on $[0, \pi)$, $h(t) = t - \pi$ on $[\pi, 2\pi)$, $h(t) = t - 2\pi$ on $[2\pi, 3\pi)$, and so on. Plot the function on the interval $[0, 30]$. It should have the appearance of a sawtooth wave.

18. This problem is based on Problem 35 in Section 6.2 of Boyce & DiPrima. Consider Bessel's equation

$$ty'' + y' + ty = 0.$$

We will use the Laplace Transform to find some leading terms of the power series expansion of the Bessel function of order zero that is continuous at the origin. Let $y(t)$ be such a solution of the equation, and suppose $Y(s)$ is its transform.

(a) Show that $Y(s)$ satisfies the equation

$$(1 + s^2)Y'(s) + sY(s) = 0.$$

(b) Solve the equation in part (a) using **DSolve**. (Use $Y(0) = 1$.)

(c) Use the **Series** command to expand the solution in part (b) in powers of $1/s$ out to terms of degree 10. (*Hint:* Do not expand around $s = 0$.)

(d) Apply the inverse Laplace Transform to part (c) to obtain the power series expansion of y up to degree 10.

Chapter 12

Higher Order Equations and Systems of First Order Equations

First and second order differential equations arise naturally in many applications. For example, first order differential equations occur in models of population growth and radioactive decay, and second order equations in the study of the motion of a falling body or the motion of a pendulum. There are several techniques for solving special classes of first and second order equations. These techniques produce symbolic solutions, expressed by a formula. For equations that cannot be solved by any of these techniques, we use numerical, geometric, or qualitative methods to investigate solutions.

Equations of higher order, and systems of first order equations, also arise naturally. For example, third order equations come up in fluid dynamics; fourth order equations, in elasticity; and systems of first order equations, in the study of spring–mass systems with several springs and masses. For general higher order equations and systems, there are hardly any techniques for obtaining explicit formula solutions, and numerical, geometric, or qualitative techniques must be used. Nevertheless, for the special class of constant coefficient linear equations and first order linear systems, there are techniques for producing explicit solutions.

In this chapter, we consider higher order equations and first order systems. We outline the basic theory in the linear case. We show how to solve linear first order systems, first by using *Mathematica* to compute eigenpairs of the coefficient matrix, and then by using **DSolve**. In addition, we discuss the plotting of phase portraits using *Mathematica*.

Higher Order Linear Equations

Consider the linear equation

$$y^{(n)} + p_1(x)y^{(n-1)} + \cdots + p_n(x)y = g(x), \tag{1}$$

and assume the coefficient functions are continuous on the interval $a < x < b$. Then the following are true:

- If $y_1(x), y_2(x), \ldots, y_n(x)$ are n linearly independent solutions of the corresponding homogeneous equation,

$$y^{(n)} + p_1(x)y^{(n-1)} + \cdots + p_n(x)y = 0, \tag{2}$$

and if y_p is any particular solution of the inhomogeneous equation (1), then the general solutions of the equations (1), (2) are, respectively,

$$y = \sum_{i=1}^{n} C_i y_i(x) + y_p(x)$$

and

$$y = \sum_{i=1}^{n} C_i y_i(x).$$

- Given a point $x_0 \in (a, b)$ and initial values $y(x_0) = y_0$, $y'(x_0) = y_0'$, $\ldots$, $y^{(n-1)}(x_0) = y_0^{(n-1)}$, there is a unique solution to equation (1) that satisfies the initial data; the solution has n continuous derivatives and exists on the entire interval (a, b).

- A set of n solutions $y_1(x), y_2(x), \ldots, y_n(x)$ of (2) is linearly independent if and only if the Wronskian

$$W(y_1, y_2, \ldots, y_n) = \det \begin{pmatrix} y_1 & y_2 & \cdots & y_n \\ y_1' & y_2' & \cdots & y_n' \\ \vdots & \vdots & \ddots & \vdots \\ y_1^{(n-1)} & y_2^{(n-1)} & \cdots & y_n^{(n-1)} \end{pmatrix}$$

is nonzero at some point in (a, b); a linearly independent collection y_1, y_2, $\ldots$, y_n is called a *fundamental set* of solutions of (2).

We see that the general theory of higher order linear equations parallels that of first and second order equations. We note, however, that solutions may be difficult to construct. In practice, we can construct fundamental sets of solutions only for constant coefficient and Euler equations. In these cases, solutions can be constructed by considering trial solutions e^{rx} and x^r, as in the second order case. There is a difficulty even for these equations if n is large, since it is difficult to compute the roots of the characteristic polynomial. We can, however, use *Mathematica* to find good approximations to the roots.

The **DSolve** command can solve many homogeneous constant coefficient and Euler equations. You can confirm this by solving several linear equations of higher order from your text.

Systems of First Order Equations

Consider the general system of first order equations,

$$
\begin{aligned}
x_1' &= F_1(t, x_1, x_2, \ldots, x_n) \\
x_2' &= F_2(t, x_1, x_2, \ldots, x_n) \\
&\vdots \\
x_n' &= F_n(t, x_1, x_2, \ldots, x_n).
\end{aligned} \tag{3}
$$

Such systems arise in the study of spring–mass systems with many springs and masses, and in many other applications. Systems of first order equations also arise in the study of higher order equations. A single higher order differential equation

$$
y^{(n)} = F(t, y, y', \ldots, y^{(n-1)})
$$

can be converted into a system of first order equations, simply by setting

$$
x_1 = y, x_2 = y', \ldots, x_n = y^{(n-1)}.
$$

The resulting system is

$$
\begin{aligned}
x_1' &= x_2 \\
x_2' &= x_3 \\
&\vdots \\
x_{n-1}' &= x_n \\
x_n' &= F(t, x_1, x_2, \ldots, x_n).
\end{aligned}
$$

It is not surprising that the problem of analyzing the general system (3) is quite formidable. Generally, the techniques we can bring to bear will be numerical, geometric, or qualitative. We shall discuss the application of these techniques to nonlinear equations in Chapter 13. For now, we merely observe that we can expect to make some progress toward a formula solution for *linear systems*. So, consider a linear system

$$
\begin{aligned}
x_1' &= a_{11}(t)x_1 + \ldots + a_{1n}(t)x_n + b_1(t) \\
x_2' &= a_{21}(t)x_1 + \ldots + a_{2n}(t)x_n + b_2(t) \\
&\vdots \quad\ \vdots \\
x_n' &= a_{n1}(t)x_1 + \ldots + a_{nn}(t)x_n + b_n(t).
\end{aligned}
$$

We can write this system in matrix notation:

$$
\mathbf{X}' = \mathbf{A}(t)\mathbf{X} + \mathbf{B}(t),
$$

where

$$\mathbf{X} = \begin{pmatrix} x_1(t) \\ x_2(t) \\ \vdots \\ x_n(t) \end{pmatrix}, \qquad \mathbf{B} = \begin{pmatrix} b_1(t) \\ b_2(t) \\ \vdots \\ b_n(t) \end{pmatrix},$$

and

$$\mathbf{A} = \begin{pmatrix} a_{11}(t) & a_{12}(t) & \cdots & a_{1n}(t) \\ \vdots & \vdots & \ddots & \vdots \\ a_{n1}(t) & a_{n2}(t) & \cdots & a_{nn}(t) \end{pmatrix}.$$

There is a general theory for linear systems, which is completely parallel to the theory for linear differential equations. If $\mathbf{X}^{(1)}, \mathbf{X}^{(2)}, \ldots, \mathbf{X}^{(n)}$ is a family of linearly independent solutions, *i.e.*, a *fundamental set* of solutions, of the homogeneous equation (with $\mathbf{B} = 0$), and if $\mathbf{X}_p$ is a particular solution of the inhomogeneous equation, then the general solutions of these equations are, respectively,

$$\mathbf{X} = \sum_{i=1}^{n} C_i \mathbf{X}^{(i)}$$

and

$$\mathbf{X} = \sum_{i=1}^{n} C_i \mathbf{X}^{(i)} + \mathbf{X}_p.$$

The test for linear independence of n solutions is

$$\det(\mathbf{X}^{(1)} \ldots \mathbf{X}^{(n)}) \neq 0.$$

For linear systems, the constant coefficient case is the easiest to handle, just as it was for single linear equations. Consider the homogeneous linear system

$$\mathbf{X}' = \mathbf{A}\mathbf{X}, \tag{4}$$

where

$$\mathbf{A} = \begin{pmatrix} a_{11} & a_{12} & \cdots & a_{1n} \\ \vdots & \vdots & \ddots & \vdots \\ a_{n1} & a_{n2} & \cdots & a_{nn} \end{pmatrix}$$

is an $n \times n$ matrix of constants. We seek solutions of the form $\mathbf{X}(t) = \boldsymbol{\xi} e^{rt}$; note the similarity to the case of a single linear equation. We find that $\mathbf{X}(t)$ is a nonzero solution of (4) if and only if

$$\mathbf{A}\boldsymbol{\xi} = r\boldsymbol{\xi},$$

i.e., if and only if r is an *eigenvalue* of $\mathbf{A}$ and $\boldsymbol{\xi}$ is a corresponding *eigenvector*. We call the pair $r, \boldsymbol{\xi}$ an *eigenpair*. In order to explain how to construct a fundamental set of solutions, we have to consider three separate cases.

CASE I. **A** has distinct real eigenvalues.

Let $r_1, r_2, \ldots, r_n$ be the distinct eigenvalues of **A**, and let $\boldsymbol{\xi}^{(1)}, \boldsymbol{\xi}^{(2)}, \ldots, \boldsymbol{\xi}^{(n)}$ be corresponding eigenvectors. Then

$$\mathbf{X}^{(1)}(t) = \boldsymbol{\xi}^{(1)} e^{r_1 t}, \ldots, \mathbf{X}^{(n)}(t) = \boldsymbol{\xi}^{(n)} e^{r_n t}$$

is a fundamental set of solutions for (4), and

$$\mathbf{X}(t) = c_1 \boldsymbol{\xi}^{(1)} e^{r_1 t} + \cdots + c_n \boldsymbol{\xi}^{(n)} e^{r_n t}$$

is the general solution.

CASE II. **A** has distinct eigenvalues, some of which are complex.

Since **A** is real, if $r, \boldsymbol{\xi}$ is a complex eigenpair, then the complex conjugate $\bar{r}, \bar{\boldsymbol{\xi}}$ is also an eigenpair. Thus, the corresponding solutions

$$\mathbf{X}^{(1)}(t) = \boldsymbol{\xi} e^{rt}, \quad \mathbf{X}^{(2)}(t) = \bar{\boldsymbol{\xi}} e^{\bar{r} t}$$

are conjugate. Therefore, we can find two real solutions of (4) corresponding to the pair $r, \bar{r}$ by taking the real and imaginary parts of $\mathbf{X}^{(1)}(t)$ or $\mathbf{X}^{(2)}(t)$. Writing $\boldsymbol{\xi} = \mathbf{a} + i\mathbf{b}$, where **a** and **b** are real, and $r = \lambda + i\mu$, where λ and μ are real, we have

$$
\begin{aligned}
\mathbf{X}^{(1)}(t) &= (\mathbf{a} + i\mathbf{b}) e^{(\lambda + i\mu)t} \\
&= (\mathbf{a} + i\mathbf{b}) e^{\lambda t} (\cos \mu t + i \sin \mu t) \\
&= e^{\lambda t} (\mathbf{a} \cos \mu t - \mathbf{b} \sin \mu t) + i e^{\lambda t} (\mathbf{a} \sin \mu t + \mathbf{b} \cos \mu t).
\end{aligned}
$$

Thus, the vector functions

$$\mathbf{u}(t) = e^{\lambda t} (\mathbf{a} \cos \mu t - \mathbf{b} \sin \mu t)$$

$$\mathbf{v}(t) = e^{\lambda t} (\mathbf{a} \sin \mu t + \mathbf{b} \cos \mu t)$$

are real solutions to (4).

To keep the discussion simple, assume that $r_1 = \lambda + i\mu$, $r_2 = \lambda - i\mu$ are complex, and that $r_3, \ldots, r_n$ are real and distinct. Let the corresponding eigenvectors be $\boldsymbol{\xi}^{(1)} = \mathbf{a} + i\mathbf{b}$, $\boldsymbol{\xi}^{(2)} = \mathbf{a} - i\mathbf{b}$, $\boldsymbol{\xi}^{(3)}, \ldots, \boldsymbol{\xi}^{(n)}$. Then

$$\mathbf{u}(t), \mathbf{v}(t), \boldsymbol{\xi}^{(3)} e^{r_3 t}, \ldots, \boldsymbol{\xi}^{(n)} e^{r_n t}$$

is a fundamental set of solutions for (4). The general situation should now be clear.

CASE III. **A** has repeated eigenvalues.

We restrict our discussion to the case $n = 2$. Suppose $r = \rho$ is an eigenvalue of **A** of multiplicity 2, meaning that ρ is a double root of the characteristic polynomial $\det(\mathbf{A} - r\mathbf{I})$. There are still two possibilities. If we can find two linearly independent eigenvectors $\boldsymbol{\xi}^{(1)}$ and $\boldsymbol{\xi}^{(2)}$ corresponding to ρ, then the solutions $\mathbf{X}^{(1)}(t) = \boldsymbol{\xi}^{(1)} e^{\rho t}$ and $\mathbf{X}^{(2)}(t) = \boldsymbol{\xi}^{(2)} e^{\rho t}$ form a fundamental set. The other possibility is that there is only one (linearly independent) eigenvector $\boldsymbol{\xi}$ with eigenvalue ρ. Then one solution of (4) is given by

$$\mathbf{X}^{(1)}(t) = \boldsymbol{\xi} e^{\rho t},$$

and a second by

$$\mathbf{X}^{(2)}(t) = \boldsymbol{\xi} t e^{\rho t} + \boldsymbol{\eta} e^{\rho t},$$

where η satisfies

$$(\mathbf{A} - r\mathbf{I})\boldsymbol{\eta} = \boldsymbol{\xi}.$$

The vector η is called a *generalized eigenvector* of $\mathbf{A}$ for the eigenvalue ρ.

Using *Mathematica* to Solve Linear Systems

We now show how to use *Mathematica* to find the eigenpairs of a matrix, and thus to find fundamental sets of solutions and the general solution of a system of linear differential equations with constant coefficients.

EXAMPLE 1. Consider the system

$$\mathbf{x}' = \begin{pmatrix} 3 & -2 & 0 \\ 2 & -2 & 0 \\ 0 & 1 & 1 \end{pmatrix} \mathbf{x}.$$

We enter the coefficient matrix in *Mathematica* as a list of lists:

```
A = {{3, -2, 0}, {2, -2, 0}, {0, 1, 1}}
```

We can display **A** in standard matrix form using the command **MatrixForm[A]**, which produces the output

$$\begin{pmatrix} 3 & -2 & 0 \\ 2 & -2 & 0 \\ 0 & 1 & 1 \end{pmatrix}$$

We find the eigenpairs of **A** with the command **Eigensystem[A]**, which produces

$$\{\{-1, 1, 2\}, \{\{-1, -2, 1\}, \{0, 0, 1\}, \{2, 1, 1\}\}\}$$

The output of this command is a list containing first a list of the eigenvalues and then a list of the corresponding eigenvectors. In this example, the eigenpairs are

$$r_1 = -1, \; \boldsymbol{\xi}^{(1)} = \begin{pmatrix} -1 \\ -2 \\ 1 \end{pmatrix}$$

$$r_2 = 1, \; \boldsymbol{\xi}^{(2)} = \begin{pmatrix} 0 \\ 0 \\ 1 \end{pmatrix}$$

$$r_3 = 2, \; \boldsymbol{\xi}^{(3)} = \begin{pmatrix} 2 \\ 1 \\ 1 \end{pmatrix}.$$

Thus, a fundamental set of solutions is

$$\mathbf{X}^{(1)}(t) = \begin{pmatrix} -1 \\ -2 \\ 1 \end{pmatrix} e^{-t}, \quad \mathbf{X}^{(2)}(t) = \begin{pmatrix} 0 \\ 0 \\ 1 \end{pmatrix} e^{t}, \quad \mathbf{X}^{(3)}(t) = \begin{pmatrix} 2 \\ 1 \\ 1 \end{pmatrix} e^{2t},$$

and the general solution is

$$\mathbf{x}(t) = c_1 \begin{pmatrix} -1 \\ -2 \\ 1 \end{pmatrix} e^{-t} + c_2 \begin{pmatrix} 0 \\ 0 \\ 1 \end{pmatrix} e^{t} + c_3 \begin{pmatrix} 2 \\ 1 \\ 1 \end{pmatrix} e^{2t}.$$

Suppose we are seeking the specific solution with initial value

$$\mathbf{x}(0) = \begin{pmatrix} 3 \\ 5 \\ 0 \end{pmatrix}.$$

The constants c_1, c_2, and c_3 must satisfy

$$c_1 \begin{pmatrix} -1 \\ -2 \\ 1 \end{pmatrix} + c_2 \begin{pmatrix} 0 \\ 0 \\ 1 \end{pmatrix} + c_3 \begin{pmatrix} 2 \\ 1 \\ 1 \end{pmatrix} = \begin{pmatrix} -1 & 0 & 2 \\ -2 & 0 & 1 \\ 1 & 1 & 1 \end{pmatrix} \begin{pmatrix} c_1 \\ c_2 \\ c_3 \end{pmatrix} = \begin{pmatrix} 3 \\ 5 \\ 0 \end{pmatrix}.$$

To solve this linear system, we use **LinearSolve**. Let **m** be the coefficient matrix and let **b** be the right-hand side:

```
m = {{-1, 0, 2}, {-2, 0, 1}, {1, 1, 1}};
b = {3, 5, 0};
LinearSolve[m, b]
```

The last command produces the output $\{-\frac{7}{3}, 2, \frac{1}{3}\}$. Thus the solution of the linear system is $c_1 = -7/3$, $c_2 = 2$, $c_3 = 1/3$, and the solution to our initial value problem is

$$\mathbf{x}(t) = -\frac{7}{3} \begin{pmatrix} -1 \\ -2 \\ 1 \end{pmatrix} e^{-t} + 2 \begin{pmatrix} 0 \\ 0 \\ 1 \end{pmatrix} e^{t} + \frac{1}{3} \begin{pmatrix} 2 \\ 1 \\ 1 \end{pmatrix} e^{2t}.$$

(Note that the columns of the matrix **m** are the eigenvectors of **A**, so an easy way to define **m** is to type **m = Transpose[Last[Eigensystem[A]]]**.)

EXAMPLE 2. Consider the system

$$\mathbf{x}' = \begin{pmatrix} 3 & -2 \\ 4 & -1 \end{pmatrix} \mathbf{x} \qquad \text{(Boyce \& DiPrima, Sect. 7.6, Prob. 1).}$$

We set **B = {{3, -2}, {4, -1}}**, and find the eigenpairs with the command **Eigensystem[B]**, which produces

$$\{\{1 - 2I, 1 + 2I\}, \{\{1 - I, 2\}, \{1 + I, 2\}\}\}$$

So the eigenpairs are

$$1 - 2i, \quad \begin{pmatrix} 1 - i \\ 2 \end{pmatrix}$$

and

$$1 + 2i, \quad \begin{pmatrix} 1 + i \\ 2 \end{pmatrix}.$$

Note that the second pair is the complex conjugate of the first. Thus a complex fundamental set is

$$\mathbf{X}^{(1)}(t) = \begin{pmatrix} 1 - i \\ 2 \end{pmatrix} e^{(1-2i)t}, \quad \mathbf{X}^{(2)}(t) = \begin{pmatrix} 1 + i \\ 2 \end{pmatrix} e^{(1+2i)t}.$$

The real and imaginary parts of the solution can be extracted by hand or by using the *Mathematica* commands **ComplexExpand, Re** and **Im** as follows:

```
ComplexExpand[Re[{1 - I, 2}* Exp[(1 - 2*I)*t]]]
ComplexExpand[Im[{1 - I, 2}* Exp[(1 - 2*I)*t]]]
```

(See the *Glossary* or the online help for more information on the above commands.) Either way, we see that a fundamental set of real solutions is

$$\mathbf{u}(t) = e^t \left[\begin{pmatrix} 1 \\ 2 \end{pmatrix} \cos 2t + \begin{pmatrix} -1 \\ 0 \end{pmatrix} \sin 2t \right]$$

$$\mathbf{v}(t) = e^t \left[- \begin{pmatrix} 1 \\ 2 \end{pmatrix} \sin 2t + \begin{pmatrix} -1 \\ 0 \end{pmatrix} \cos 2t \right].$$

EXAMPLE 3. Consider the system

$$\mathbf{x}' = \begin{pmatrix} 3 & -4 \\ 1 & -1 \end{pmatrix} \mathbf{x} \qquad \text{(Boyce \& DiPrima, Sect. 7.7, Prob. 1)}.$$

We set **A = {{3, -4}, {1, -1}}**. The command **Eigensystem[A]** produces

$$\{\{1, 1\}, \{\{2, 1\}, \{0, 0\}\}\}$$

The number 1 is listed twice as an eigenvalue, and $\{0, 0\}$ is listed as an eigenvector (although it really isn't one because eigenvectors must be nonzero). This is *Mathematica's* way of reporting that 1 is an eigenvalue of multiplicity 2 and that $\{2, 1\}$ is the only corresponding eigenvector. Thus $\rho = 1$ and $\boldsymbol{\xi} = \begin{pmatrix} 2 \\ 1 \end{pmatrix}$, and

$$\mathbf{X}^{(1)}(t) = \begin{pmatrix} 2 \\ 1 \end{pmatrix} e^t$$

is one solution. To find a second solution we solve

$$(\mathbf{A} - \rho \mathbf{I})\boldsymbol{\eta} = \begin{pmatrix} 2 & -4 \\ 1 & -2 \end{pmatrix} \boldsymbol{\eta} = \boldsymbol{\xi} = \begin{pmatrix} 2 \\ 1 \end{pmatrix}.$$

Let **m = {{2, -4}, {1, -2}}** be the coefficient matrix, and let **b = {2, 1}** be the right-hand side. Then **LinearSolve[m, b]** produces $\{1, 0\}$, so

$$\eta = \begin{pmatrix} 1 \\ 0 \end{pmatrix}.$$

(This system actually has an infinite number of solutions; **LinearSolve** returns one of them.) Thus, a second solution is

$$\mathbf{X}^{(2)}(t) = \xi t e^{\rho t} + \eta e^{\rho t} = \begin{pmatrix} 2 \\ 1 \end{pmatrix} t e^t + \begin{pmatrix} 1 \\ 0 \end{pmatrix} e^t.$$

The general solution is

$$\mathbf{X}(t) = c_1 \begin{pmatrix} 2 \\ 1 \end{pmatrix} e^t + c_2 \left[\begin{pmatrix} 2 \\ 1 \end{pmatrix} t e^t + \begin{pmatrix} 1 \\ 0 \end{pmatrix} e^t \right].$$

We have shown how to use *Mathematica*'s linear algebra commands to find eigenpairs, and thus solutions, for linear systems of differential equations with constant coefficients. The solutions to systems of linear equations can also be found with a direct application of **DSolve**.

EXAMPLE 4. Consider the system $x' = y$, $y' = -x$. To find its general solution, you would type

DSolve[{x'[t] == y[t], y'[t] == -x[t]}, {x[t], y[t]}, t]

The result is

$$x(t) = -c_1 \cos t + c_2 \sin t,$$

$$y(t) = c_2 \cos t + c_1 \sin t.$$

To solve the same system with initial conditions $x(0) = 1$, $y(0) = 0$, you would type

DSolve[{x'[t] == y[t], y'[t] == -x[t], x[0] == 1,
** y[0] == 0}, {x[t], y[t]}, t]**

Both methods for solving linear systems are useful. Solving in terms of the eigenpairs of the coefficient matrix is somewhat involved, but yields a simple and useful formula for the solution. The eigenpairs contain valuable information about the solution, which we will exploit further in Problem Set F. On the other hand, it is simpler to use **DSolve**, and for many purposes the solutions generated by **DSolve** are completely satisfactory.

Here are some examples on which you can practice:

(a) $x' = -3x + 2y$, $y' = -x$
(b) $x' = x - 2y$, $y' = -x$
(c) $x' = x + 2y$, $y' = -x$.

Phase Portraits

A solution of a 2×2 linear system is a pair of funtions $x(t)$, $y(t)$. The plot of a solution as a function of t would be a curve in (t, x, y)-space. Generally, such 3-dimensional plots are too complicated to be illuminating. Therefore, we project the curve from 3-dimensional space into the (x, y)-plane. This plane is called the *phase plane*, and the resulting curve in the phase plane is called a *trajectory*. A trajectory is drawn by plotting $(x(t), y(t))$ as the parameter t varies. A plot of a family of trajectories is called a *phase portrait* of the system.

A trajectory is an example of a *parametrized curve*. The basic *Mathematica* command for plotting a parametrized curve is **ParametricPlot**. In this section, we describe how to use *Mathematica*: (1) to solve an initial value problem consisting of a linear system and an initial condition, and plot the corresponding trajectory; (2) to solve a linear system using **DSolve** and draw a phase portrait; (3) to solve a system using **NDSolve** and draw a phase portrait.

Plotting a Single Trajectory

Consider the linear system in Example (a) above, with initial condition $x(0) = 1$, $y(0) = 0$. We want to plot the trajectory with this initial condition. First we enter the initial value problem.

```
ivp = {x'[t] == -3x[t] + 2y[t], y'[t] == -x[t], x[0] == 1,
   y[0] == 0}
```

Then we type the following:

```
solivp[t_] =
  {x[t], y[t]} /. First[DSolve[ivp, {x[t], y[t]}, t]]
traj[t0_, t1_] := ParametricPlot[Evaluate[solivp[t]],
  {t, t0, t1}, AspectRatio -> Automatic]
```

This program does two things. First, it produces the solution, **solivp**, of the initial value problem; this solution can be evaluated or plotted. Second, it defines a function **traj** that plots the trajectory corresponding to **solivp**. The function **traj** takes two arguments, $t0$ and $t1$, the endpoints of the time interval for the trajectory. Figure 1 contains the picture generated by typing **traj[-5, 5]**. To use this program for plotting the solution of a different initial value problem, all you have to do is modify the definition of **ivp** appropriately.

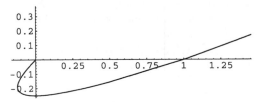

Figure 1

Plotting Several Trajectories

Consider the linear system in Example (b) above. We want to plot trajectories of this system with various initial conditions $x(0) = a$, $y(0) = b$. We first define a function **sol** that solves this system with initial conditions $x(0) = a$, $y(0) = b$.

```
ivp = {x'[t] == x[t] - 2y[t], y'[t] == -x[t], x[0] == a,
   y[0] == b}
sol[t_, a_, b_] =
  {x[t], y[t]} /. First[DSolve[ivp, {x[t], y[t]}, t]]
```

The function **sol** can be evaluated for a particular choice of initial conditions $x(0) = a$, $y(0) = b$, at a particular value of t.

Next we define a function **phase** that plots a phase portrait of the linear system. We do this by creating a table of solutions with different initial conditions, and then using **ParametricPlot** to plot this table of solution functions.

```
phase[t0_, t1_] :=
  ParametricPlot[Evaluate[Flatten[Table[sol[t, a, b],
      {a, -2, 2}, {b, -2, 2}], 1]], {t, t0, t1},
   AspectRatio -> Automatic]
```

The **Table** command embedded in this definition produces a list with two levels of braces; we used the **Flatten[..., 1]** command to remove the unnecessary braces. Note that **sol** was defined with **=**, but **:=** was used in the definition of **phase** to suppress plotting until a time interval is specified. We have chosen a rectangular grid of 25 initial conditions, with a ranging from -2 to 2 and b ranging from -2 to 2 (in integer increments).

To use the function **phase**, simply enter **phase[t0, t1]**, where $t0$ and $t1$ are the endpoints of the time interval $[t0, t1]$ for which you wish to plot the trajectories. This will plot 25 trajectories corresponding to the 25 different initial conditions. The time interval $[t0, t1]$ must be chosen somewhat carefully; a small time interval could result in solution curves that are too short to get an idea of where they're headed, while a large time interval could cause the scale of the graph to get very large, resulting in solution curves that are squeezed together. Moreover, the larger the time interval, the longer it will take for *Mathematica* to produce the plot. You should start with a small time interval like $[-3, 3]$ and then adjust the interval to get a satisfactory picture. The command **phase[-3, 3]** generated the phase portrait shown in Figure 2.

You should be able to plot the phase portraits for any 2×2 linear system with constant coefficients using the procedure above, replacing the differential equations in the definition of **ivp**, and choosing, by trial and error, an appropriate time interval to use with the function **phase**. By combining the phase portrait with a plot of the vector field (see Chapter 13), you should be able to determine the direction of increasing t on the curves. Also, in each case you should be able to decide whether the origin is a *sink* (all solutions approach the origin as t increases),

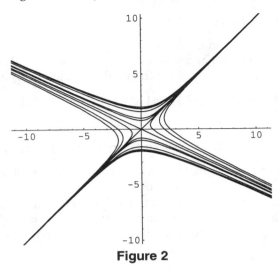

Figure 2

source (all solutions move away from the origin), *saddle point* (solutions approach the origin along one direction, then move away from the origin in another direction), or *center* (solutions follow closed curves around the origin). The procedure for plotting a phase portrait must be modified slightly if you wish to vary only one parameter in the initial data. For example, the program

```
ivp = {x'[t] == x[t] + 2y[t], y'[t] == -x[t], x[0] == a,
   y[0] == 0}
sol[t_, a_] =
   {x[t], y[t]} /. First[DSolve[ivp, {x[t], y[t]}, t]]
phase[t0_, t1_] :=
   ParametricPlot[Evaluate[Table[sol[t, a], {a, -4, 4}]],
   {t, t0, t1}, AspectRatio -> Automatic]
```

will yield (upon entering **phase[-3, 3]**) the nine curves shown in Figure 3 as a phase portrait for Example (c). (Count the curves! One of the solutions is the equilibrium solution $x(t) = 0$, $y(t) = 0$.)

Plotting Trajectories of Solutions Generated by NDSolve

The symbolic solver **DSolve** should be able to solve any homogeneous 2×2 linear system with constant coefficients. For inhomogeneous linear systems, linear systems with variable coefficients, and nonlinear systems, **DSolve** will generally not do the job. In these situations, we turn to the numerical solver **NDSolve**.

A naive approach to using **NDSolve** would be to simply replace **DSolve** with **NDSolve** in the programs above. This will not work, in part because the syntax of **NDSolve** is different from that of **DSolve**, but more importantly, because **NDSolve**

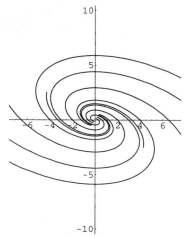

Figure 3

must have numerical values for the initial conditions before it is evaluated. To circumvent this problem you can define the numerical solution using delayed evaluation (**:=**). Here is an example that uses **NDSolve** on the linear system in Example (b) above.

```
nsol[tval_, a_, b_] :=
  ({x[t], y[t]} /. First[NDSolve[{x'[t] == x[t] - 2y[t],
      y'[t] == -x[t], x[0] == a, y[0] == b},
      {x[t], y[t]}, {t, -3, 3}]]) /. t -> tval
```

Then you could incorporate the function **nsol** into the **phase** function defined above (replacing **sol** by **nsol**), and then enter **phase[-3, 3]** to plot the trajectories.

Note that in the definition of **nsol**, the time interval on which **NDSolve** will compute the solution is built into the definition (in this case, **{t, -3, 3}**), and therefore must be fixed in advance. A more flexible scheme, useful for some of the problems in Problem Set F, is given by

```
nsol[tval_, a_, b_, t1_] :=
  ({x[t], y[t]} /. First[NDSolve[{x'[t] == x[t] - 2y[t],
      y'[t] == -x[t], x[0] == a, y[0] == b},
      {x[t], y[t]}, {t, 0, t1}]]) /. t -> tval
nphase[t1_] :=
  ParametricPlot[Evaluate[Flatten[Table[nsol[t, a, b, t1],
    {a, -3, 3}, {b, -3, 3}], 1]],{t, 0, t1},
  AspectRatio -> Automatic]
```

This program starts the solution curves at time $t = 0$ and gives you the freedom to choose the final time. You could, of course, define a slightly more complicated function that would allow you to vary both endpoints of the time interval.

Chapter 13

Qualitative Theory for Systems of Differential Equations

In this chapter, we extend the qualitative theory of autonomous differential equations from single equations to systems of two equations. Thus, consider a system of the form

$$\begin{cases} x' = F(x, y) \\ y' = G(x, y). \end{cases} \tag{1}$$

We assume the functions F and G have continuous partial derivatives everywhere in the plane, and the critical points—the common zeros of F and G—are isolated.

We know from the fundamental existence and uniqueness theorems that, corresponding to any choice of initial data (x_0, y_0), there is a unique pair of functions $x(t)$, $y(t)$ that satisfy the system (1) of differential equations and the initial conditions $x(0) = x_0$, $y(0) = y_0$. Taken together, this pair of functions parametrizes a curve, or trajectory, in the *phase plane* that passes through the point (x_0, y_0). Since we are dealing with autonomous systems (*i.e.*, the variable t does not appear on the right side of (1)), the solution curves are *independent of the starting time*. That is, if we consider initial conditions $x(t_0) = x_0$, $y(t_0) = y_0$, then we get the same solution curve with a time delay of t_0 units. Moreover, we also know that two distinct trajectories cannot intersect. Thus, the plane is covered by the family of trajectories; the plot of these curves is the *phase portrait* of (1).

As we know, we can find explicit formula solutions $x(t)$, $y(t)$ only for simple systems. Thus, we are forced to turn to qualitative and numerical methods. It is our purpose here to discuss qualitative techniques for studying the solutions of (1).

In most cases of physical interest, every solution curve behaves in one of the following ways:

  (i) The curve consists of a single critical point;

  (ii) The curve tends to a critical point as $t \to \infty$;

  (iii) The curve is unbounded: the distance from $(x(t), y(t))$ to the origin becomes arbitrarily large as $t \to \infty$, or as $t \to t^*$, for some finite t^*;

  (iv) The curve is periodic: the parametric functions satisfy $x(t + t_p) = x(t)$, $y(t + t_p) = y(t)$ for some fixed t_p;

  (v) The curve approaches a periodic solution, *e.g.*, spiraling in on a circle.

Qualitative techniques may be able to identify the kinds of solutions that appear. In order to keep matters simple, we try to answer the following specific questions:

(1) What can we say about the critical points? Can we make a qualitative guess about their nature and stability? Can we deduce anything about the solution curves that start out close to the critical points?

(2) Can we predict anything about the long-term behavior of the solution curves? This includes those near critical points as well as the solution curves in general.

The list of possible limiting behaviors of solution curves is based on the Poincaré-Bendixson Theorem (*cf.* Theorem 9.7.3 in Boyce & DiPrima), which is valid for autonomous systems of two equations. Systems of three or more equations can have much more complicated limiting behavior. For example, solution curves can remain bounded and yet fail to approach any equilibrium or periodic state as $t \to \infty$. This phenomenon, called "chaos" by mathematicians, was anticipated by Poincaré (and perhaps even earlier by the physicist Maxwell), but most scientists did not appreciate how widespread chaos is until the arrival of computers. See Problem 13 in Problem Set F for an example.

There are two qualitative methods for analyzing systems of equations: one based on the idea of a *vector field*; and the other based on *linearized stability analysis*. The latter method is treated in detail in Sections 9.3–9.5 of Boyce & DiPrima; it provides information about the stability of critical points of a nonlinear system by studying associated linear systems. In this chapter, we focus on the former method: the use of vector fields in qualitative analysis. Some of the problems in Problem Set F involve linearized stability analysis.

Here is the basic idea. For any solution curve $(x(t), y(t))$, and any point t, we have the differential equations

$$\begin{cases} x'(t) = F(x(t), y(t)) \\ y'(t) = G(x(t), y(t)). \end{cases}$$

If we employ vector notation

$$\mathbf{x}(t) = \begin{pmatrix} x(t) \\ y(t) \end{pmatrix}, \quad \mathbf{f}(\mathbf{x}) = \begin{pmatrix} F(\mathbf{x}) \\ G(\mathbf{x}) \end{pmatrix},$$

then the system is written

$$\mathbf{x}' = \mathbf{f}(\mathbf{x}).$$

Moreover, the vector $\mathbf{x}'(t) = \mathbf{f}(\mathbf{x}(t))$ is just the tangent vector to the curve $\mathbf{x}(t)$. Since knowledge of the collection of tangent vectors to the solution curves would give a good idea of the curves themselves, it would be useful to have a plot of these vectors. Thus, corresponding to any vector $\mathbf{x}$, we draw the vector $\mathbf{f}(\mathbf{x})$, translated so that its foot is at the point $\mathbf{x}$. This plot is called the vector field of the system (1). Since we cannot do this at every point of the phase plane, we only draw the vectors at a set of regularly chosen points. Then we step back and look at the resulting vector field. We do not have the solution curves sketched, but we do have a representative collection of their tangent vectors. From the tangent vectors, we can get a good idea of the curves themselves. Indeed, we can often answer the questions above by a careful examination of the vector field.

Vector fields can be drawn by hand for simple systems, but the command **PlotVectorField**, which you have been using to plot direction fields of single first order differential equations, can draw vector fields of any first order system of two equations. We illustrate the use of this command by examining two specific systems, namely

$$\begin{cases} x' = x(1 - x - y) \\ y' = y(0.75 - 0.5x - y); \end{cases} \tag{2}$$

and

$$\begin{cases} x' = x(5 - x - y) \\ y' = y(-2 + x). \end{cases} \tag{3}$$

System (2) is a competing species model that is discussed in Boyce & DiPrima in Section 9.4, Example 1. System (3) is a predator-prey model from Section 7.4 of Edwards & Penney, **Elementary Differential Equations with Applications**, 3rd edition.

The **PlotVectorField** command is in the **Graphics`PlotField`** package, and is loaded by typing <<**Graphics`PlotField`**. The syntax is

```
PlotVectorField[{F[x, y], G[x, y]}, {x, x0, x1},
  {y, y0, y1}]
```

We first consider system (2). To draw the vector field in Figure 1, we enter

```
PlotVectorField[{x(1 - x - y), y(0.75 - 0.5x - y)},
  {x, 0, 1.5}, {y, 0, 1.5}, Frame -> True]
```

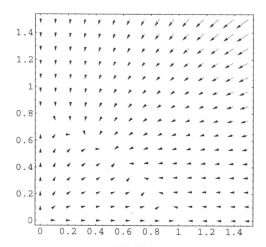

Figure 1

In a vector field, the vector drawn at the point $\mathbf{x} = (x, y)$ indicates both the direction and length of $\mathbf{f}(\mathbf{x})$. The vector at $\mathbf{x}$ drawn by **PlotVectorField** faithfully represents the direction of $\mathbf{f}(\mathbf{x})$, but the length of each vector is rescaled so that the longest vector is the same size as the distance between neighboring points. The critical, or equilibrium, points are those points at which the vector field vanishes, *i.e.*, those points where $F(x, y) = G(x, y) = 0$. However, it is difficult to tell exactly where the critical points are. Since most of the vectors are so short that we see only their arrowheads, we cannot tell which ones are really shortest. Nonetheless, we can approximately locate the critical points as places where the direction of the arrows is changing rapidly.

We can improve the picture by changing the relative sizes of the vectors. One way to do this would be to make the vectors a constant length. This can be done with the option **ScaleFunction** $->$**(1&)**. The "&" indicates to *Mathematica* that the constant "1" is to be interpreted as a pure function. A better way is to use a scaling function that lengthens the vectors near the critical points enough that their directions are clearly indicated, but not so much that the critical points cannot be identified. For example, we can use an option of the form **ScaleFunction** $->$**(# + c &)** for some constant c. The symbol "#" refers to the argument of the scaling function. Thus a vector of length proportional to r is replaced by a vector of length proportional to $r + c$. The constant c can be found by trial and error; the idea is to make the tails of the arrows visible, but short, near the critical points. Increasing c lengthens the shorter arrows, but a value of c that is too large will make all the arrows nearly the same length. Ideally, c should be larger than, but resonably close to, the maximum value of the length of the vector field in the given plotting rectangle.

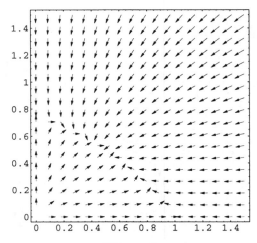

Figure 2

Here is the command we use to plot the "improved" vector field. We find that a value of $c = 5$ works well in this case.

```
PlotVectorField[{x(1 - x - y), y(0.75 - 0.5x - y)},
  {x, 0, 1.5}, {y, 0, 1.5}, Frame -> True,
  ScaleFunction -> (# + 5 &), PlotPoints -> 18]
```

The option **PlotPoints** $\rightarrow$ **18** increases slightly the number of vectors plotted. The result is shown in Figure 2. This picture gives us a much clearer indication of what's going on, but it still doesn't allow us to identify the critical points precisely. To do this we actually have to solve the equations $x' = 0$ and $y' = 0$ simultaneously. We can do this with the **Solve** command,

```
Solve[{x(1 - x - y) == 0, y(0.75 - 0.5x - y) == 0},
  {x, y}]
```

This produces the answer

$$\{\{x \rightarrow 0., y \rightarrow 0.\}, \{x \rightarrow 0., y \rightarrow 0.75\}, \{x \rightarrow 0.5, y \rightarrow 0.5\}, \{x \rightarrow 1., y \rightarrow 0.\}\}$$

Thus, the critical points are $(0, 0)$, $(0, 0.75)$, $(1, 0)$, and $(0.5, 0.5)$. Knowing the critical points, we can deduce from the vector field that $(0, 0)$ is an unstable node, $(0, 0.75)$ and $(1, 0)$ are unstable saddle points, and $(0.5, 0.5)$ is an asymptotically stable node. Solutions starting near the saddle points (but not on the axes) tend away from them, and those starting near the point $(0.5, 0.5)$ tend toward it. In fact, the vector field strongly suggests that every solution curve starting in the first quadrant (but not on the axes) tends toward $(0.5, 0.5)$. Hence $(0.5, 0.5)$, which corresponds to equal populations, is apparently the limiting state that all positive solutions approach as t increases. We have thus answered both questions (1) and (2) on the basis of the vector field. Figure 3 provides a closer look at the vector field near the point $(0.5, 0.5)$.

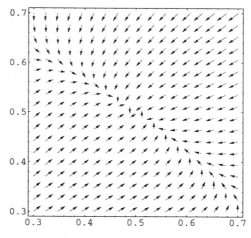

Figure 3

Now consider the predator-prey system (3). Here $x(t)$ represents the population of prey and $y(t)$ the population of predators. The critical points of this system are $(0,0), (5,0), (2,3)$. As in system (2), our goal is to use the vector field to answer questions (1) and (2). Here is the appropriate *Mathematica* command:

```
PlotVectorField[{x(5 - x - y),y(-2 + x)}, {x, 0, 6},
    {y, 0, 5}, Frame -> True, ScaleFunction -> (# + 75 &),
    PlotPoints -> 18]
```

This time we used the constant 75 in the scaling function. The output is shown in Figure 4.

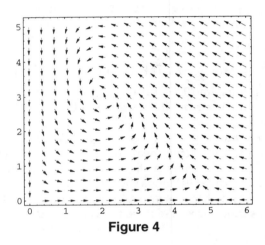

Figure 4

It appears from the portrait that $(0,0)$ and $(5,0)$ are unstable saddle points and that solutions spiral around $(2,3)$. The vectors near $(2,3)$ suggest that the solutions spiral into $(2,3)$, and linearized stability analysis confirms that $(2,3)$ is an asymptotically stable spiral point. In particular, solutions starting near the first two critical points tend away from them, whereas solution curves starting near the latter tend toward it. In fact, the vector field suggests strongly that every solution curve starting in the first quadrant (but not on the axes) tends toward the critical point $(2,3)$. This is the strongest kind of stable equilibrium—namely, no matter what the initial populations are, as long as they are both positive, the population tends toward the equilibrium position of 40% predators and 60% prey.

Remarks

- We indicated above that vector fields could be drawn by hand. It would be difficult, however, to draw as satisfactory a vector field as we have produced with *Mathematica*. On the other hand, we note that a careful study of the signs of F and G can lead to a good understanding of the nature of the vector field.

This is done for the system of Example (2) in Section 9.4 in Boyce & DiPrima. See in particular Figure 9.4.5 there.

- In Chapter 6 we used a command of the form

```
PlotVectorField[{1, h[x, y]}, {x, x0, x1}, {y, y0, y1}]
```

to plot the direction field of the single first order differential equation

$$\frac{dy}{dx} = h(x, y).$$

This corresponds to plotting the vector field of the system

$$\begin{cases} x' = 1 \\ y' = h(x, y). \end{cases}$$

The "1" here explains the **1** that was used in the command for plotting direction fields.

- The essential idea in Chapter 6 for a single autonomous first order equation

$$x' = f(x)$$

is to determine the qualitative nature of the solutions purely on the basis of the sign of $f(x)$ for various x's. Actually, we have done the same thing here: namely, determine the qualitative nature of the solutions (trajectories) of a system from information on the signs of F and G.

- As mentioned above, according to linearized stability analysis, the nature of critical points for nonlinear systems can be deduced from the associated linear systems. If a critical point is asymptotically stable for the associated linear system (the case if the eigenvalues are negative or have negative real parts), then the point is asymptotically stable for the original nonlinear system. If it is unstable for the linear system, it is unstable for the nonlinear system. Centers are ambiguous. This analysis is valid in exactly the same way for a single equation. Suppose $\bar{x}$ is a critical point for $x' = f(x)$. Then letting $u(t) = x(t) - \bar{x}$, we see that

$$u' = f(u + \bar{x}) \approx f(\bar{x}) + f'(\bar{x})u = f'(\bar{x})u,$$

so the associated linear equation is $u' = f'(\bar{x})u$. The solutions are $u(t) = ce^{\lambda t}$, where $\lambda = f'(\bar{x})$. Therefore, $x(t) = \bar{x} + ce^{\lambda t}$. It is clear that $\bar{x}$ is asymptotically stable if $\lambda < 0$ and unstable if $\lambda > 0$. The same is true of the nonlinear system (with the situation for $\lambda = 0$ being ambiguous). Thus, we see that this assessment of stability parallels the situation for systems.

Conclusion

You can calculate solutions of nonlinear systems with **NDSolve** and plot their trajectories. We gave suggestions on convenient ways to do this in Chapter 12. Taken together, the information provided by a vector field, by a plot of the trajectories, and by linearized stability analysis, gives a complete understanding of the behavior of the solutions of a system of two first order equations. The information provided by these three approaches is similar, but each approach yields information the other approaches don't provide. For example, a vector field provides a good global indication of what the phase portrait will look like, as well as indicating the approximate location of the critical points and their type; a plot of the trajectories gives a precise indication of the solution curves; and a linearized stability analysis determines the type of the critical points.

In using these approaches to study a system, it is almost always best to start with a vector field plot or with linearized stability analysis. Typically, it takes a long time to plot trajectories of nonlinear systems. But by using the information from a vector field plot or from linearized stability analysis, you can choose appropriate initial data points to use with **NDSolve**. A vector field plot will also help you choose an appropriate time interval (*e.g.*, positive or negative time).

Problem Set F

Systems of Equations

The solution to Problem 6 appears in the *Sample Notebook Solutions*. Also, in Chapter 12 you can find a number of suggestions for using *Mathematica* to solve these problems. The parameters in those suggestions may need to be modified according to the problem at hand. In particular, you may have to change the time interval or the set of initial conditions.

1. In this problem, we study three systems of equations taken from Boyce & DiPrima:

$$\mathbf{x}' = \begin{pmatrix} 2 & -1 \\ 3 & -2 \end{pmatrix} \mathbf{x} \qquad \text{(Prob. 3, Sect. 7.5)} \qquad\qquad (i)$$

$$\mathbf{x}' = \begin{pmatrix} 1 & -1 \\ 5 & -3 \end{pmatrix} \mathbf{x} \qquad \text{(Prob. 5, Sect. 7.6)} \qquad\qquad (ii)$$

$$\mathbf{x}' = \begin{pmatrix} -3 & 5/2 \\ -5/2 & 2 \end{pmatrix} \mathbf{x} \qquad \text{(Prob. 4, Sect. 7.7).} \qquad\qquad (iii)$$

   (a) Use *Mathematica* to find the eigenvalues and eigenvectors of each linear system. Use the eigenvectors and eigenvalues to write down the general solution of the system.

   (b) Use **DSolve** to compute the solution of system (i). Compare the general solution you obtained in part (a) with the solution generated by **DSolve**. Are they the same? If not, explain how they are related (that is, match up the constants).

   (c) Using the solution formulas obtained above, plot several trajectories of each system. On your graphs, identify the eigenvectors (if relevant), the direction of increasing time on the trajectories, and the type and stability of the origin as a critical point. You may find that the quality of your portraits can be enhanced by modifying the t-interval or the range of values assumed by the initial data. (A word of caution: If you request too many curves, *Mathematica* will take a long time to generate the plot.)

2. Do Problem 1 but replace the three systems listed there by the following three from Boyce & DiPrima:

$$\mathbf{x}' = \begin{pmatrix} 5/4 & 3/4 \\ 3/4 & 5/4 \end{pmatrix} \mathbf{x} \qquad \text{(Prob. 6, Sect. 7.5)} \qquad (i)$$

$$\mathbf{x}' = \begin{pmatrix} 2 & -5/2 \\ 9/5 & -1 \end{pmatrix} \mathbf{x} \qquad \text{(Prob. 4, Sect. 7.6)} \qquad (ii)$$

$$\mathbf{x}' = \begin{pmatrix} -3/2 & 1 \\ -1/4 & -1/2 \end{pmatrix} \mathbf{x} \qquad \text{(Prob. 3, Sect. 7.7).} \qquad (iii)$$

3. Use **Eigensystem** to find the eigenvalues and eigenvectors of the following systems. Use the eigenvalues and eigenvectors to write down the general solutions.

(a)

$$\mathbf{x}' = \begin{pmatrix} 2 & -1 \\ 1 & -2 \end{pmatrix} \mathbf{x}.$$

Let $\mathbf{x}(t) = (x(t), y(t))$. Determine the possible limiting behavior of $x(t)$, of $y(t)$, and of $x(t)/y(t)$ as t approaches $+\infty$.

(b)

$$\mathbf{x}' = \begin{pmatrix} 4 & 3 \\ -3 & -2 \end{pmatrix} \mathbf{x}.$$

(c)

$$\mathbf{x}' = \begin{pmatrix} 0 & 0 & -1 \\ 2 & 0 & 0 \\ -1 & 2 & 4 \end{pmatrix} \mathbf{x}.$$

Find the solution with initial condition

$$\mathbf{x}(0) = \begin{pmatrix} 7 \\ 5 \\ 5 \end{pmatrix}.$$

(d) Now solve the initial value problem in (c) with **DSolve** and compare the answer with the solution you obtained in (c).

4. *Mathematica* can solve some inhomogeneous linear systems. Here are three such systems (taken from Boyce & DiPrima, Sect. 7.9, Problems 1, 3, & 10).

$$\mathbf{x}' = \begin{pmatrix} 2 & -1 \\ 3 & -2 \end{pmatrix} \mathbf{x} + \begin{pmatrix} e^t \\ t \end{pmatrix} \qquad (i)$$

$$\mathbf{x}' = \begin{pmatrix} 2 & -5 \\ 1 & -2 \end{pmatrix} \mathbf{x} + \begin{pmatrix} -\cos t \\ \sin t \end{pmatrix} \qquad (ii)$$

$$\mathbf{x}' = \begin{pmatrix} -3 & \sqrt{2} \\ \sqrt{2} & -2 \end{pmatrix} \mathbf{x} + \begin{pmatrix} e^{-t} \\ -e^{-t} \end{pmatrix}. \tag{iii}$$

(a) Solve the three systems using **DSolve**.

(b) Now impose the initial data

$$\mathbf{x}(0) = \begin{pmatrix} 1 \\ 1 \end{pmatrix}$$

on each system. For each of (*i*)–(*iii*), draw the solution curve using a time interval $-1 \le t \le 1$. Based on the differential equation, indicate the direction of motion through the initial value. Then expand the time interval and, purely on the graphical evidence, venture a guess as to the behavior of the solution curve as $t \to -\infty$ and as $t \to \infty$.

5. Here we reconsider the pendulum models examined in Problems 3–6 of Problem Set D. (See also the discussion in Section 9.3 of Boyce & DiPrima.)

   (a) Consider first the undamped pendulum

$$\theta'' + \sin \theta = 0, \quad \theta(0) = 0, \quad \theta'(0) = b.$$

Let $x = \theta$ and $y = \theta'$; then x and y satisfy the system

$$\begin{cases} x' = y \\ y' = -\sin x \end{cases} \qquad \begin{cases} x(0) = 0 \\ y(0) = b. \end{cases}$$

Use **NDSolve** to solve this system. Then plot, on a single graph, the resulting trajectories for the initial conditions $x(0) = 0$, and $y(0) = 0.5, 1, 1.5, 2, 2.5$. Use a positive time range (*e.g.*, **{t, 0, 20}**) and the option **AspectRatio -> Automatic** to improve your pictures.

   (b) Based on your pictures in part (a), describe physically what the pendulum seems to be doing in the three cases $\theta'(0) = 1.5, 2$, and 2.5.

   (c) The energy of the pendulum is defined as the sum of kinetic energy $(\theta')^2/2$ and potential energy $1 - \cos \theta$, so

$$E = \frac{1}{2}(\theta')^2 + 1 - \cos \theta = \frac{1}{2}y^2 + 1 - \cos x.$$

Show, by taking the derivative dE/dt, that E is constant when $\theta(t)$ is a solution to the pendulum equation. What basic physical principle does this represent?

   (d) Use **ContourPlot** (with the options **ContourShading -> False, PlotPoints -> 60, AspectRatio -> Automatic, AxesOrigin -> {0, 0}**) to plot the level curves for the energy. Explain how your picture is related to the trajectory plot from part (a).

   (e) If the pendulum reaches the upright position (*i.e.*, $\theta = \pi$), what must be true of the energy? Now explain why there is a critical value E_0 of the energy, below which the pendulum swings back and forth without reaching the upright position, and above which it swings overhead and continues to revolve in the same direction. What is E_0? What must the pendulum do when $E = E_0$? Explain. Next, consider the solution curve corresponding to initial data $\theta(0) = 0, \theta'(0) = b$. What is the value of the energy on that curve? Do your conclusions from part (b) agree with your analysis in this part?

(f) Now consider the damped pendulum

$$\theta'' + 0.5\theta' + \sin\theta = 0, \quad \theta(0) = 0, \quad \theta'(0) = b.$$

Use **NDSolve** to solve the corresponding first order system (which you must determine) and plot the resulting trajectories for the initial velocities $b = 0, 0.5, 1, \ldots, 6$. You will notice that all the trajectories in your graph tend toward the critical points $(0,0)$ or $(2\pi, 0)$. Explain what this means physically. There is a value b_0 for which the trajectory tends toward the critical point $(\pi, 0)$. Estimate, up to two-decimal accuracy, the value of b_0. What would this correspond to physically?

(g) Recall the energy defined in part (c). Compute E' and determine which of the following are possible: (1) E is increasing, (2) E is decreasing, (3) E stays the same. Explain how the possibilities are reflected in the solutions from part (f).

6. Consider the *competing species* model (Boyce & DiPrima, Prob. 2, Sect. 9.4)

$$\begin{cases} \dfrac{dx}{dt} = x(1.5 - x - 0.5y) \\ \dfrac{dy}{dt} = y(2 - 1.5x - 0.5y). \end{cases}$$

(a) Find all critical points of the system. At each critical point, calculate the corresponding linear system and find the eigenvalues of the coefficient matrix; then identify the type and stability of the critical point.

(b) Plot the vector field.

(c) Use several initial data points (x_0, y_0) in the first quadrant to draw a phase portrait for the system. Identify the direction of increasing t on the trajectories you obtain. Use your vector field plot from part (b) to choose a representative sample of initial conditions. Then combine the vector field and phase portrait on a single graph.

(d) Suppose the initial state of the population is given by

$$x(0) = 2.5, \ y(0) = 2.$$

Find the state of the population at $t = 1, 2, 3, 4, 5, \ldots, 20$.

(e) Explain why practically speaking there is no "peaceful coexistence"; *i.e.*, with the exception of an atypical set of starting populations (the *separatrix* curves), one or the other population must die out. For which nonzero initial populations is there no change? Sketch (by hand) on your plot the separatrices; *i.e.,* the solution curves that approach the unstable equilibrium point where both populations are positive and form the boundary between the solution curves that approach each of the two stable points where only one population survives. (In Problems 11 and 12, we investigate how to approximate separatrices using *Mathematica*.)

(f) The vertical line $x = 1.5$ cuts the separatrix. By experimenting with the long-term behavior of solutions emanating from several points on this line, find an approximation to two-decimal point accuracy of the value $\bar{y}$ such that $(1.5, \bar{y})$ is on the separatrix. Could you verify numerically whether the point is in fact on the separatrix? Why or why not?

7. Consider the *competing species* model (Boyce & DiPrima, Prob. 4, Sect. 9.4)

$$\begin{cases} \dfrac{dx}{dt} = x(1.5 - 0.5x - y) \\ \dfrac{dy}{dt} = y(0.75 - 0.125x - y). \end{cases}$$

(a) Find all critical points of the system. At each critical point, calculate the corresponding linear system and find the eigenvalues of the coefficient matrix; then identify the type and stability of the critical point. (*Note:* In Version 2.2 of *Mathematica*, you must enter the decimal numbers as fractions in the **Solve** command.)

(b) Plot the vector field.

(c) Use several initial data points (x_0, y_0) in the first quadrant to draw a phase portrait for the system. Identify the direction of increasing t on the trajectories you obtain. Use your vector field plot from part (b) to choose a representative sample of initial conditions. Then combine the vector field and phase portrait on a single graph.

(d) Suppose the initial state of the population is given by

$$x(0) = 0.1, \ y(0) = 0.1.$$

Find the state of the population at $t = 1, 2, 3, 4, 5, \ldots, 20$.

(e) Explain why, practically speaking, "peaceful coexistence" is the only outcome; *i.e.,* with the exception of the situation in which one or both species starts out without any population, the population distributions always tend toward a certain equilibrium point. Sketch (by hand) on your plot the separatrices that connect the stable equilibrium point to the two unstable points at which one population is zero; these separatrices divide the solution curves that tend toward the origin as $t \to -\infty$ from those that are unbounded as $t \to -\infty$. (In Problems 11 and 12, we investigate how to approximate separatrices using *Mathematica*.)

(f) The vertical line $x = 2.5$ cuts a separatrix. By experimenting with the behavior of solutions for *negative* time emanating from several points on this line, find an approximation to two-decimal point accuracy of the value $\bar{y}$ such that $(2.5, \bar{y})$ is on the separatrix. (*Hint:* $-3 \leq t \leq 0$ should suffice.) Could you verify numerically whether the point is in fact on the separatrix? Why or why not?

8. Consider the *predator-prey* model

$$\begin{cases} \dfrac{dx}{dt} = x(2 - y) \\ \dfrac{dy}{dt} = y(x - 1). \end{cases}$$

in which x represents the population of the prey and y represents the population of the predators.

(a) Find all critical points of the system. At each critical point, calculate the corresponding linear system and find the eigenvalues of the coefficient matrix; then identify the type and stability of the critical point.

(b) Plot the vector field.

(c) Use several initial data points (x_0, y_0) in the first quadrant to draw a phase portrait for the system. Identify the direction of increasing t on the trajectories you obtain. Use your vector field plot from part (b) to choose a representative sample of initial conditions. Then combine the vector field and phase portrait on a single graph.

(d) Explain from your phase portrait how the populations vary for initial data close to the equilibrium point $(\bar{x}_0, \bar{y}_0)$ (the unique critical point inside the first quadrant). What about initial data far from the critical point?

(e) Suppose the initial state of the population is given by

$$x(0) = 1, \; y(0) = 1.$$

Find the state of the population at $t = 1, 2, 3, 4, 5$.

(f) Estimate the period of the solution curve with initial data $(1, 1)$.

9. Consider a *predator-prey* model where the behavior of the prey is governed by a logistic equation (in the absence of the predator). Such a model is typified by the following system (Boyce & DiPrima, Prob. 3, Sect. 9.5):

$$\begin{cases} \dfrac{dx}{dt} = x(1 - 0.5x - 0.5y) \\ \dfrac{dy}{dt} = y(-0.25 + 0.5x). \end{cases}$$

where x represents the population of the prey, and y represents the population of the predators.

(a) Find all critical points of the system. At each critical point, calculate the corresponding linear system and find the eigenvalues of the coefficient matrix; then identify the type and stability of the critical point.

(b) Plot the vector field.

(c) Use several initial data points (x_0, y_0) in the first quadrant to draw a phase portrait for the system. Identify the direction of increasing t on the trajectories you obtain. Use your vector field plot from part (b) to choose a representative sample of initial conditions. Then combine the vector field and phase portrait on a single graph.

(d) Explain from your phase portrait how the populations vary over time for initial data near the equilibrium point $(\bar{x}_0, \bar{y}_0)$ (the unique critical point inside the first quadrant, not on the axes). What happens when the initial condition is far from the equilibrium point?

(e) Suppose the initial state of the population is given by

$$x(0) = 1, \; y(0) = 1.$$

Find the state of the population at $t = 1, 2, 3, 4, 5$.

(f) Estimate how long it takes for both populations to arrive simultaneously within 0.01 of their equilibrium values if we start with initial data $(1, 1)$.

10. Consider a modified predator-prey system where the behavior of the prey is governed by a logistic/threshold equation (in the absence of the predator). Such a model is typified by the following system (Boyce & DiPrima, Prob. 5, Sect. 9.5):

$$\begin{cases} \dfrac{dx}{dt} = x(-1 + 2.5x - 0.3y - x^2) \\[2mm] \dfrac{dy}{dt} = y(-1.5 + x). \end{cases}$$

where x represents the population of the prey, and y represents the population of the predators.

(a) Find all critical points of the system. At each critical point, calculate the corresponding linear system and find the eigenvalues and eigenvectors of the coefficient matrix; then identify the critical points as to type and stability.

(b) Plot the vector field.

(c) Use several initial data points (x_0, y_0) in the first quadrant to draw a phase portrait for the system. Use your vector field plot from part (b) to choose a representative sample of initial conditions.

(d) In parts (a), (b), and (c), you obtained information about the solutions of the system using three different approaches. Combine all of this information by combining the plots from parts (b) and (c) with the **Show** command, and then drawing in by hand the critical points, the eigenvectors from part (a), and any separatrices. What information is provided by the approach in part (a) that is not provided by the other approaches? Answer the same question for parts (b) and (c).

(e) Interpret your conclusions in terms of the populations of the two species.

11. Some of the nonlinear systems of differential equations we have studied have more than one asymptotically stable equilibrium point. In such cases, it can be difficult to predict which (if any) equilibrium point the solution with a given initial condition will approach as time increases. Generally, the solution curves that approach one stable equilibrium will be separated from the solutions that approach another stable equilibrium by a separatrix curve. The separatrix is itself a solution curve that does not approach either stable equilibrium—often it approaches a saddle point instead. (See, for example, Figure 9.4.4 in Section 9.4 of Boyce & DiPrima.)

Having located a relevant saddle point, one can approximate the separatrix by choosing an initial condition very close to the saddle point and solving the differential equation *backwards* in time. The reason for going "back in time" is to find (approximately) a solution curve that approaches very close to the saddle point as time increases. For two population dynamics models, we will use this idea to get a fairly precise picture of which initial conditions go to which equilibria.

(a) Consider the system

$$\begin{cases} \dfrac{dx}{dt} = x(1.5 - x - 0.5y) \\ \dfrac{dy}{dt} = y(2 - 1.5x - 0.5y). \end{cases}$$

studied in Problem 6 above. There are two asymptotically stable equilibria, $(1.5, 0)$ and $(0, 4)$, a saddle point at $(1, 1)$, and an unstable node at $(0, 0)$. Plot a family of solution curves in the first quadrant (positive x and y) on the same graph; use a time interval from $t = 0$ to a positive time large enough that you can clearly see where the solutions curves are headed, but not so large that your plot takes forever to compute. (We suggest a set of initial values between 0.5 and 4.5. It also may be useful to adjust the range of the plot until you get a reasonably square graph.) Observe where the solution curves seem to be heading and where the separatrix seems to be.

(b) Draw an approximate separatrix by plotting a solution curve with initial values for x and y very close to the saddle point (1,1), using a *negative* range of values for t. This should give you a good picture of the separatrix on one side of the saddle point; to approximate the rest of the separatrix you will need to choose a set of initial conditions on the other side of (but still very close to) the saddle point. In each case, you may need to fine-tune the time interval, since a time interval that is too small will only show you a small part of the separatrix, while too large an interval may cause the solution to blow up. Finally, show all of your plots on one graph, and make sure the separatrix really does separate the different asymptotic properties of the solution curves found in part (a). (To help in distinguishing the separatrix, you can use the option **PlotStyle** -> **GrayLevel[.5]** in making your separatrix plots.) Discuss how the possible limiting values of the species depend on the initial values of the species.

12. Read the introduction to Problem 11.

(a) Do part (a) of Problem 11, but using the system

$$\begin{cases} \dfrac{dx}{dt} = x(-1 + 2.5x - 0.3y - x^2) \\ \dfrac{dy}{dt} = y(-1.5 + x). \end{cases}$$

from Problem 10 above. This time there are asymptotically stable equilibria at the origin and $(3/2, 5/3)$, and saddle points at $(0.5, 0)$ and $(2, 0)$.

(b) Do part (b) of Problem 11 for the system in part (a) of this problem. It is up to you to decide whether one or both of the saddle points are relevant to drawing the separatrix or separatrices that divide those solution curves that approach one equilibrium from those that approach the other as time increases; your picture from part (a) may be helpful. Also, keep in mind that since the saddle points are on the x-axis in this case, you need only consider nearby initial conditions above the x-axis because we are only interested in solutions in the first quadrant.

13. This problem is based on Problems 7.3.34–35 of Edwards and Penney, **Elementary Differential Equations with Applications**, 3rd edition.

(a) Consider the linear system

$$\begin{cases} \dfrac{dx}{dt} = -x + hy \\ \dfrac{dy}{dt} = x - y. \end{cases}$$

Draw phase portraits for the system for the values $h = 0, 0.5$, and -0.5. What can you deduce about the change in the type and/or stability of the equilibrium point $(0, 0)$ corresponding to relatively small perturbations of the system?

(b) Now consider the system

$$\begin{cases} \dfrac{dx}{dt} = y + hx(x^2 + y^2) \\ \dfrac{dy}{dt} = -x + hy(x^2 + y^2). \end{cases}$$

Draw phase portraits (using **NDSolve** this time) for $h = 0, 1$, and -1. (*Note:* For $h = 1$, you may have to keep the initial data close to the origin and keep the time interval very short. Even with an upper limit $t = 0.5$, *Mathematica* complains; but be patient, you will get a picture eventually. The picture may contain extraneous lines, caused by *Mathematica* trying to extrapolate beyond singularities of some of the solutions. If you want to suppress these lines, see the *Complex Systems* section below for a hint.) Illustrate how linearized stability analysis is inconclusive in the case of a *center* by citing the evidence in your portraits. (To decide which way the solution curves are going, you can examine the signs of dx/dt and dy/dt at a particular point, say $x = 1$, $y = 0$, or use *Mathematica* to plot the vector field in each case.)

14. In Chapter 5 we investigated the sensitivity to initial values of solutions of a single differential equation. In this problem, we investigate the same issue for two systems of equations. The systems we study are said to be *chaotic* because the solutions are sensitive to initial values, but in contrast to the examples in Chapter 5 and Problems 14 and 15 of Problem Set C, the solutions remain bounded.

Since we do not know the exact solutions for the systems we will study, in order to judge the time it takes for a small perturbation of the system to become large, we will compare pairs of numerical solutions whose initial conditions are close, but still relatively far apart compared with the local error (*cf.* Chapter 7) of our numerical method. Since **NDSolve** tries to keep its errors within 10^{-6}, to be safe we only consider perturbations as small as 10^{-4}.

   (a) Consider the Lorenz system

$$x' = 10(y - x)$$
$$y' = 28x - y - xz$$
$$z' = -(8/3)z + xy$$

(which is studied, for example, in Section 9.8 of Boyce & DiPrima). We investigate the idea that small changes in the initial conditions can lead to large changes in the solution after a relatively short period of time. For $a = 0.1$, plot on the same graph the first coordinate $x(t)$ of the solutions corresponding to the initial conditions $(5, 5, 5)$ and $(5 + a, 5, 5)$ from $t = 0$ to $t = 20$. (It may be useful in distinguishing between the two curves to make one of them black and the other gray—you can do this with a **Plot** option like **PlotStyle -> {{}, GrayLevel[0.5]}**. You may also have to increase **PlotPoints**.) Observe the time at which the solutions start to differ noticeably. Repeat for $a = 0.01, 0.001, 0.0001$. (The criterion for a "noticeable" difference between the two solutions is up to you; just try to be consistent from one observation to the next.)

   (b) Make a table or graph of the number of decimal places in which the initial condition was perturbed (that is, 1, 2, 3, and 4 for the four given values of a) versus the amount of time the solutions stayed close to each other. This graph shows roughly how long we should trust a numerical solution to be close to the actual solution for a given number of digits of accuracy in each step of our numerical method. Describe how the amount of time we can trust a numerical solution seems to depend on the number of digits of accuracy per time step, judging from your data. If a numerical method has 16 digits of accuracy, about how long do you think the numerical solution can be trusted? What if the accuracy were 100 digits?

15. Read the introduction to Problem 14.

(a) Do part (a) of Problem 14, but using the Rössler system

$$x' = -y - z$$
$$y' = x + 0.36y$$
$$z' = 0.4x - 4.5z + xz.$$

This time plot from $t = 0$ to $t = 100$ and use $(2, 2, 2)$ as the initial condition in place of $(5, 5, 5)$.

(b) Do part (b) of Problem 14 for the Rössler system.

Complex Systems

In the following three problems, we consider a special class of 2×2 systems, namely, those that may be represented as a differential equation involving a single complex variable z. Consider, for example, the equation

$$z' = z(e^{i\phi} + z\bar{z}). \tag{1}$$

Here $z(t) = x(t) + iy(t)$ is a complex-valued function of the independent variable t, ϕ is a real parameter, and $\bar{z}$ is the complex conjugate of z. If we compute the real and imaginary parts of (1), we obtain the equivalent nonlinear 2×2 system

$$\begin{cases} x' = x \cos \phi - y \sin \phi + x^3 - xy^2 \\ y' = x \sin \phi + y \cos \phi + x^2 y - y^3. \end{cases} \tag{2}$$

In spite of the fact that these are equivalent, the expression (1) offers several advantages over (2). First, (1) is "one-dimensional" in the sense that there is only one dependent variable z. Second, there are certain properties of the equation (1) that are apparent in the complex form, but obscure in the 2×2 form. For example, (1) is invariant by rotations through any angle (see the following problem), and therefore its set of solutions is rotationally symmetric. Symmetry plays a large role in the theory of differential equations, and complex-valued equations like (1) are the simplest kind of differential equations displaying rotational symmetry. Third, equations like (1) exhibit a phenomenon called *bifurcation* when the parameter ϕ varies. A parametrized family of differential equations is said to *bifurcate* at a parameter value ϕ_0 if the qualitative behavior of the differential equation changes as the parameter passes through ϕ_0. Examples of bifurcations are: loss of stability of a critical point; and creation or annihilation of limit cycles, or periodic solutions.

You may have noticed that **DSolve** returns complex-valued solutions for certain real-valued differential equations. In fact, many *Mathematica* commands do not distinguish between real and complex numbers. This is true of the commands **DSolve** and **NDSolve**. In particular, we can use these commands on equations like (1) without worrying about the fact that these equations are complex-valued. *Mathematica* will report its solutions in exactly the same form as for real-valued differential equations, except that now the solutions will almost always be complex-valued (whereas in the case of real-valued equations, complex-valued solutions were rarely reported).

Unfortunately, the *Mathematica* command **ParametricPlot** *does* distinguish between complex- and real-valued functions, so if we want to draw the phase portraits of the solutions in the complex plane, we first have to extract the real and imaginary parts of the complex solutions, and then use **ParametricPlot** on the real and imaginary parts. In order to save you a considerable amount of trouble figuring out exactly how to do this, we give you the following program:

```
complexsystem[phi_, t0_, t1_] := (Clear[csol, cvar, g];
  csol[t_, r_, theta_] := z[t] /.
    NDSolve[{z'[t] == z[t]*(Exp[I*phi]
        + z[t]*Conjugate[z[t]]), z[0] == r*Exp[I*theta]},
      z[t], {t, t0, t1}];
  coords[cvar_] = If[Abs[cvar] > 3, {},
  {Re[cvar], Im[cvar]}];
  g[t_] = Map[coords, Flatten[Table[csol[t, 0.25i,
      2Pi*j/6], {i, 1, 3, 2}, {j, 0, 5}]]];
  ParametricPlot[Evaluate[g[t]], {t, t0, t1},
    PlotRange -> {{-2, 2}, {-2, 2}},
    AspectRatio -> Automatic];)
```

This program is designed to solve equation (1) numerically and plot solutions for a given value of the parameter ϕ. The other arguments of the function are the endpoints of the time interval on which to solve the differential equation. For example, typing **complexsystem[Pi, -5, 5]** will plot a phase portrait for $\phi = \pi$ using values of t from -5 to 5.

In order to use and modify this program, you should note the following features:

- The differential equation appears in the **NDSolve** command in exactly the same way that an ordinary real-valued equation would. Note that **I** is *Mathematica*'s notation for $i = \sqrt{-1}$.
- We have included an initial condition **z[0] == r*Exp[I*theta]**, specifying a complex number given in polar coordinates. Note that multiplication by $e^{i\theta}$ corresponds in the complex plane to counterclockwise rotation by the angle θ. The idea is to use a set of initial points which, like the differential equation, are rotationally symmetric around the origin (see parts (a)

of the problems below). In this particular example, the initial conditions consist of points on the circles of radius 0.25 and 0.75, spaced evenly at angles $2\pi j/6$, for $j = 0, \ldots, 5$.

- The function **coords** first checks to see if its complex-valued argument is outside the circle of radius 3, and thereby outside the square specified by **PlotRange**. If so, the point is thrown away, while if the point is inside the circle, then it is separated into its real and imaginary parts. The reason for throwing away points outside the plotting region is to avoid problems with *Mathematica* trying to extrapolate beyond singularities of some of the solutions.

- The *Mathematica* command **Map** is used to apply the function **coords** to the table of complex-valued numerical solutions of the differential equation. The end result is a table of real and imaginary parts of the complex solutions.

- When you evaluate the program, you will encounter a deluge of warning messages from *Mathematica*. Nonetheless, the correct graph should appear. Once you have determined that the program is working correctly, you may want to turn these messages off (see Chapter 8). For example, you might type **Off[NDSolve::ndsz]** to turn off warning messages about the step size approaching zero. At the very least, you should delete the warning messages from the Notebook before you print it.

To summarize, the only things you'll have to modify to use this program in the following problems are:

- the actual differential equation inside the **NDSolve** command,
- the list of initial conditions inside the **Table** command, and
- possibly the parameters of the **PlotRange** option, and, if so, the cutoff radius specified in the **If** statement.

16. This exercise concerns the equation mentioned above,

$$z' = z(e^{i\phi} + z\bar{z}). \tag{i}$$

(a) Show that equation (i) is rotationally invariant in the following sense: If we let $w = e^{i\theta}z$, for any angle θ, then w satisfies the same differential equation as z; that is, $w' = w(e^{i\phi} + w\bar{w})$. You may show this by hand or by using *Mathematica* (but it's easier to do it by hand).

(b) Plot the trajectories of the differential equation for parameter values $\phi = 0, \pi/4, \pi/3, 2\pi/3$. Determine the direction of the trajectories by evaluating the right-hand side at various values of z. (We suggest plotting the trajectories on the time interval $[-3, 2]$.) What kind of symmetries can you see in the plots?

(c) Examine the results of (b). Can you detect whether the parameter has passed through a bifurcation value? What exactly is the qualitative change in the behavior of the solutions?

(d) Plot the trajectories of the system for the two parameter values $\phi = \pi/2 \pm 0.2$. (You should also change the time interval to $[-5, 3]$.) Estimate the bifurcation value of the parameter.

(e) Let $\rho = z\bar{z}$. Note that ρ is the square of the distance from the origin to z. In particular, ρ is a real-valued positive function of t. Moreover, one can show that ρ satisfies the differential equation $\rho' = 2\rho(\mathrm{Re}(e^{i\phi}) + \rho)$. Note that a critical point of this differential equation corresponds to a solution of (i) that stays a fixed distance from the origin. By finding the critical points of the differential equation for ρ, explain the results of (c) and (d).

17. Consider the complex-valued differential equation

$$z' = e^{i\phi}z + \bar{z}^2. \qquad (i)$$

Refer to the discussion preceding Problem 16 for general remarks concerning complex-valued differential equations.

(a) Show that equation (i) is rotationally invariant through an angle of $2\pi/3$ in the following sense: If we let $w = e^{2\pi i/3}z$, then w also satisfies the differential equation (i), i.e., $w' = e^{i\phi}w + \bar{w}^2$. You may show this by hand or by using *Mathematica*.

(b) Plot the trajectories of the differential equation for parameter values $\phi = 0, \pi/4, \pi/2$. You should use a set of initial conditions consisting of 12 evenly spaced points on each of the circles of radius 0.5 and 1.5. Determine the direction of the trajectories by evaluating the right-hand side at various values of z. (We suggest plotting the trajectories on the time interval $[-3, 2]$.) What symmetries can you detect in the plot? (*Note*: These plots may take several minutes to draw.)

(c) Look at the results of (b). On which interval does the parameter pass through a bifurcation value? What exactly is the qualitative change in the behavior of the solutions?

(d) Set $z = x + iy$ in equation (i), expand the equation in terms of x and y, and take the real and imaginary parts to obtain an equivalent real-valued 2×2 system.

(e) Using the result of (d), for each of the parameter values $0, \pi/4, \pi/2$, find the critical points of the system (there are four in each case). Indicate these critical points on your plots in (b).

18. Consider the complex-valued differential equation

$$z' = 0.3e^{i\phi}z - (1 + i)z^3\bar{z}^2 + \bar{z}^4. \qquad (i)$$

Refer to the discussion preceding Problem 16 for general remarks concerning complex-valued differential equations.

(a) Show that equation (*i*) is rotationally invariant through an angle of $2\pi/5$ in the following sense: If we let $w = e^{2\pi i/5}z$, then w also satisfies the differential equation (*i*).

(b) Plot the trajectories of the differential equation for parameter values $\phi = 0, \pi/4, \pi/2$. You should use a set of initial conditions consisting of 20 evenly spaced points on each of the circles of radius 0.25 and 0.75. Determine the direction of the trajectories by plotting the vector field associated with this equation. (We suggest plotting the trajectories on the time interval $[-8, 10]$, with **PlotRange** –> **{-1,1}, {-1,1}}**.) What symmetries can you detect in the plots? (*Note:* These plots may take several minutes to draw.)

(c) Examine the results of (b). On which interval or intervals does the parameter pass through a bifurcation value? What exactly is the qualitative change in the behavior of the solutions? Using the **Show** command, plot the vector fields and integral curves on the same graph for each parameter value. By examining these plots, try to figure out how many critical points there are inside the unit square. (*Note:* The stable critical points are easy to find. The unstable ones are not so obvious.)

Glossary

We list here the *Mathematica* objects that arise in the various input statements encountered in this book. We list them according to four types: commands, options, built-in functions, and constants. The distinction between the first and third is somewhat artificial, as *Mathematica* makes no distinction between them. However, for the purposes of this book, it is convenient to think of a *Mathematica* built-in function as we normally think of functions, in particular as something that can be evaluated or plotted, while a command is something that manipulates data or expressions, or initiates a process.

Each command and option is followed by a brief description of its effect, and then one or more examples. To find the full syntax of a command you can type **?command** or **??command** in a Notebook, or use the the Help Browser.

Mathematica Commands

Clear Clears values and definitions for variables and functions.
```
Clear[x]
Clear[f, g]
```

Conjugate Gives the complex conjugate of a complex number.
```
Conjugate[2 + 3I]
```

ContourPlot Plots the level curves of a function of two variables.
```
ContourPlot[x^2 + y^2, {x, -2, 2}, {y, -2, 2}]
```

D Differentiation operator.
```
D[f[x], x]
D[x^3 - 3x, x]
D[Cos[x], {x, 2}]
```

DSolve Symbolic ODE solver.
```
DSolve[y''[x] - x*y[x] == 0, y[x], x]
DSolve[{y'[x] + y[x]^2 == 0, y[0] == 1}, y[x], x]
DSolve[{x'[t] == 2x[t] + y[t], y'[t] == -x[t]},
    {x[t], y[t]}, t]
```

Eigensystem   Computes eigenvalues and eigenvectors of a matrix.
  `Eigensystem[{{a, b}, {c, d}}]`
  `Eigensystem[{{1, 0, 0}, {1, 1, 1}, {1, 2, 4}}]`

Evaluate   Causes an expression to be evaluated; especially useful in **Plot** commands.
  `Plot[Evaluate[Table[i*x, {i, 0, 10}]], {x, 0, 1}]`

Expand   Expands products and powers in an algebraic expression.
  `Expand[(x + y)^3]`

Factor   Factors a polynomial.
  `Factor[x^4 - y^4]`

FindRoot   Finds a numerical solution to an equation; looks for a single solution near a given starting point.
  `FindRoot[Cos[x] == x, {x, 1}]`

First   Selects the first element in a list.
  `First[{2, 4, 6}]`
  `First[Solve[x^2 - 4 == 0, x]]`

Flatten   Removes extra levels of braces in a nested list.
  `Flatten[{{1, 2}, 3, {4, 5, 6, {7}}}, 1]`
  `Flatten[Table[f[a, b], {a, 1, 2}, {b, 1, 3}]]`

Function   Specifies a pure function.
  `f = Function[x, 3x^3 - x]`

If   Introduces a conditional.
  `f[x_] := If[x < 0, 4x, -4x]`

Integrate   Integration operator for both definite and indefinite integrals.
  `Integrate[1/(1 + x^2), x]`
  `Integrate[Exp[-x], {x, 0, Infinity}]`

InverseLaplaceTransform   Computes the inverse Laplace Transform.
  `InverseLaplaceTransform[Exp[-s]/s, s, t]`

Join   Combines two lists.
  `Join[{1, 2, 3}, {4, 5}]`

LaplaceTransform   Computes the Laplace Transform.
  `LaplaceTransform[f[t], t, s]`

Last   Selects the last element in a list.
  `Last[{3, 4, 5}]`

Limit   Finds the one-sided limit. By default, the limit is taken from the right, except for limits at infinity. Use **Direction->1** to take the limit from the left.
  `Limit[Sin[x]/x, x -> 0]`
  `Limit[Log[x]/x, x -> Infinity]`

LinearSolve   Solves a matrix equation.
   `LinearSolve[{{1, 2}, {3, 4}}, {6, 7}]`

ListPlot   Plots a list of points.
   `ListPlot[{{1, 1}, {2, 4}, {3, 10}}]`

Map   Applies a function to each element of a list.
   `Map[Sqrt, {1, 2, 3}]`

N   Evaluates numerically; gives a decimal approximation of a number.
   `N[Pi, 15]`
   `N[Solve[x^2 - 4x + 9 == 0, x]]`

NDSolve   Numerical ODE solver.
   `NDSolve[{y''[x] - x*y[x] == 0, y[0] == 1, y'[0] == 0},`
      `y[x], {x, 0, 10}]`

NestList   Applies a function iteratively.
   `NestList[f, x, 3]`

NIntegrate   Numerical integration.
   `NIntegrate[f[x], {x, 0, 5}]`
   `NIntegrate[Sin[x]/x, {x, 0, 1}]`

NSolve   Solves an equation (or list of equations) numerically; equivalent to
typing **N[Solve[...]]**.
   `NSolve[x^5 - x^2 + 4x - 9 == 0, x]`

Off   Suspends printing of an error or a warning message.
   `Off[NDSolve::mxst]`
   `Off[General::spell1]`

On   Turns on printing of an error or a warning message.
   `On[NDSolve::mxst]`

ParametricPlot   Draws a curve (or curves) given by parametric coordinates.
   `ParametricPlot[{Cos[t], Sin[t]}, {t, 0, 2Pi}]`
   `ParametricPlot[Evaluate[Table[f[t, j], {j, 10}]],`
      `{t, 0, 5}]`

Plot   Basic plot command for an expression (or list of expressions) of one variable.
   `Plot[f[x], {x, 0, 1}]`
   `Plot[{x^2, x^3}, {x, -2, 2}]`

PlotVectorField   Draws a vector field plot given a pair of functions of two
variables.
   `PlotVectorField[{y, -x}, {x, -1, 1}, {y, -1, 1}]`
   `PlotVectorField[{f[x, y], g[x, y]}, {x, 0, 2}, {y, 0, 5},`
      `AspectRatio -> 1]`

Remove   Removes definitions; useful for removing definitions of names accidentally invoked before a package is loaded.
    `Remove[PlotVectorField]`

Series   Generates terms of a power series expansion about a point.
    `Series[Exp[-x], {x, 0, 10}]`

Show   Combines and displays several previously generated graphics.
    `Show[graph1, graph2]`

Simplify   Simplifies an expression.
    `Simplify[1/(1 + x)^2 - 1/(1 - x)^2]`

Solve   Solves an equation or list of equations.
    `Solve[2x^2 - 3x + 6 == 0, x]`
    `Solve[{x + 3y == 4, -x - 5y == 3}, {x, y}]`

Sum   Sums a sequence.
    `Sum[(i - 1)/(i + 1), {i, 1, 10}]`
    `Sum[g[j], {j, 2, 3, 0.1}]`

Table   Generates a list.
    `Table[1/k, {k, 1, 5}]`
    `Table[{f[k], g[k]}, {k, 0, -100, -1}]`

TableForm   Places data in tabular form.
    `Table[{3t, 6t^2}, {t, 1, 50}]//TableForm`

Options to *Mathematica* Commands

AccuracyGoal   Specifies absolute error in a numerical calculation.
    `NDSolve[{y'[x] == x/y[x], y[0] == 1}, y[x], {x, 0, 2},`
        `AccuracyGoal -> 15, PrecisionGoal -> 15,`
        `WorkingPrecision -> 25]`

AspectRatio   Specifies ratio of height to width in a plot; to ensure that the same
scales are used on both axes, set **AspectRatio -> Automatic**.
    `ParametricPlot[{Cos[t], Sin[t]}, {t, 0, 2Pi},`
        `AspectRatio -> Automatic]`

Axes   Specifies whether axes should be drawn in a plot.
    `Plot[g[x], {x, 0, 3}, Axes -> False]`

AxesLabel   Specifies labels for axes in a plot.
    `Plot[Cos[x], {x, 0, 2Pi}, AxesLabel -> {"x", "y"}]`

Contours   Specifies which contour to use in a contour plot.
    `ContourPlot[f[x, y], {x, 0, 1}, {y, 0, 2},`
        `Contours -> {f[0, 0]}]`

ContourShading   Specifies whether regions between contours are shaded.
```
ContourPlot[x*y, {x, 0, 3}, {y, 0, 3},
    ContourShading -> False]
```

Dashing   Indicates that curves or lines should be drawn dashed.
```
Plot[{Sin[x], Cos[x]}, {x, 0, 2Pi},
    PlotStyle -> {{}, Dashing[{0.03}]}]
```

Direction   Specifies direction in a one-sided limit. **Direction->1** specifies the limit from the left, and **Direction->-1** specifies the limit from the right.
```
Limit[1/x, x -> 0, Direction -> 1]
```

Frame   Specifies whether a frame should be drawn around a plot. Useful for direction field plots.
```
PlotVectorField[{f[x, y], g[x, y]}, {x, -1, 1},
    {y, -2, 2}, Frame -> True]
```

GrayLevel   Allows for shading of curves in a plot.
```
Plot[{Sin[x], Cos[x]}, {x, 0, 2Pi},
    PlotStyle -> {{}, GrayLevel[0.7]}]
```

MaxIterations   Specifies number of iterations in **FindRoot**; default is 15.
```
FindRoot[-100 + 400 Exp[0.04(t - 1960)] == 3000, {t, 0},
    MaxIterations -> 40]
```

MaxSteps   Specifies number of steps allowed in **NDSolve**; default is 1000.
```
NDSolve[{y''[x] - y'[x] + x*y[x] == 0, y[0] == 0,
    y'[0] == 1}, y[x], {x, 0, 40}, MaxSteps -> 2000]
```

PlotJoined   Specifies whether to connect the points in a **ListPlot**.
```
ListPlot[{{1, 3}, {2, 5}}, PlotJoined -> True]
```

PlotPoints   Specifies the number of points to sample in drawing a plot; the default is 25 for **Plot**.
```
Plot[Sin[1/x], {x, 0.001, 1}, PlotPoints -> 50]
```

PlotRange   Specifies the range of the dependent variable in a plot.
```
Plot[Tan[x], {x, 0, Pi}, PlotRange -> {0, 10}]
ParametricPlot[{t^3, t^5}, {t, -10, 10},
    PlotRange -> {{-5, 5}, {-10, 10}}]
```

PlotStyle   Specifies style of curves or points; *cf.* **GrayLevel** or **Dashing**.

PrecisionGoal   Specifies relative error in a numerical calculation; *cf.* **AccuracyGoal**.

ScaleFunction   Option for scaling vectors in vector field plots; the argument is a pure function.
```
PlotVectorField[{f[x, y], g[x, y]}, {x, -1, 1},
    {y, -2, 2}, ScaleFunction -> (1&)]
```

Ticks   Specifies tick marks for the axes in a plot; in the following example, we set the ticks on the x-axis and allow *Mathematica* to set them on the y-axis.

```
Plot[Sin[x], {x, 0, Pi},
    Ticks -> {{0, Pi/4, Pi/2, 3Pi/4, Pi}, Automatic}]
```

WorkingPrecision   Specifies number of digits used in a calculation; *cf.* **AccuracyGoal**.

Built-in Functions

Abs $|x|$.

AiryAi   The solution Ai(x) to Airy's equation $y'' - xy = 0$.

AiryBi   The solution Bi(x) to Airy's equation.

ArcSin   arcsin x.

BesselJ   Bessel function of the first kind. **BesselJ[n, x]** is a solution, $J_n(x)$, to Bessel's equation of order n, $y'' + \frac{1}{x}y' + \left(1 - \frac{n^2}{x^2}\right)y = 0$.

BesselY   Bessel function of the second kind. **BesselY[n, x]** is also a solution, $Y_n(x)$, to Bessel's equation of order n.

Cos   cos x.

Cosh   cosh x.

Erf   The *error function* erf$(x) = (2/\sqrt{\pi}) \int_0^x e^{-t^2}\, dt$.

Erfi $-i\,\mathrm{erf}(ix)$.

Exp e^x.

Im   Im(z), the imaginary part of a complex number.

Log   The natural logarithm $\ln x = \log_e x$.

Re   Re(z), the real part of a complex number.

Sin   sin x.

Sinh   sinh x.

SinIntegral   The *sine integral* Si$(x) = \int_0^x (\sin t/t)\, dt$.

Sqrt $\sqrt{x}$.

Tan   tan x.

Built-in Constants

E $e = \ln^{-1}(1)$.

I $i = \sqrt{-1}$.

Infinity ∞.

Pi π.

Sample Notebook Solutions

These Sample Solutions were prepared as *Mathematica* 3.0 Notebooks and were printed directly from *Mathematica*.

■ Problem Set B, Problem 1

We are interested in the differential equation $x\,y' + 2\,y = \sin x,$ with initial condition $y(\pi/2) = c.$

■ (a)

Here is the general solution.

```
Clear[y, x, c, sol]
sol =
 DSolve[{x * y'[x] + 2 y[x] == Sin[x], y[Pi / 2] == c}, y[x], x]
```

$$\left\{\left\{y[x] \rightarrow \frac{-1 + \frac{c\,\pi^2}{4} - x\,\text{Cos}[x] + \text{Sin}[x]}{x^2}\right\}\right\}$$

Here is a function of two variables x and c that gives, for fixed c, the solution of the IVP with initial condition $y(\pi/2) = c.$

```
y[x_, c_] = y[x] /. First[sol]
```

$$\frac{-1 + \frac{c\,\pi^2}{4} - x\,\text{Cos}[x] + \text{Sin}[x]}{x^2}$$

The solution with initial condition $y(\pi/2) = 1$ is given by:

```
y[x, 1]
```

$$\frac{-1 + \frac{\pi^2}{4} - x\,\text{Cos}[x] + \text{Sin}[x]}{x^2}$$

■ (b)

Here is a table of values for the solution of the IVP with $c = 1$.

```
Table[{x, N[y[x, 1]]}, {x, 0.5, 5, 0.5}] // TableForm
```

```
0.5        6.03214
1.         1.76857
1.5        1.04835
2.         0.802248
2.5        0.650997
3.         0.508722
3.5        0.358712
4.         0.207823
4.5        0.0710347
5.         -0.0363934
```

■ (c)

We graph the solution **y[x, 1]** on several intervals.

```
Plot[y[x, 1], {x, 0, 2}];
```

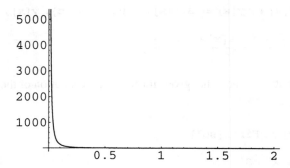

Plot[y[x, 1], {x, 1, 10}, PlotRange -> {-0.15, 1}];

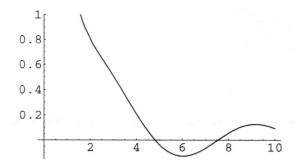

Plot[y[x, 1], {x, 10, 100}];

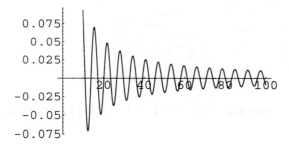

By inspecting the function **y[x, 1]**, we see that the solution approaches positive ∞ as x goes to 0 and approaches 0 as x goes to infinity. This is borne out by the graphs above.

■ **(d)**

Here are a table of the solutions of the IVP with initial conditions $y(\pi/2) = 0.2\,j$, for $j = 1, \ldots, 5$, and a plot of these solutions over two different intervals. Note that the index $j = 1$ corresponds to the lowest solution, and the index $j = 5$ corresponds to the topmost.

```
Table[{j, y[x, 0.2 * j]}, {j, 1, 5}] // TableForm
```

1     $\dfrac{-0.50652 - x \, \mathrm{Cos}[x] + \mathrm{Sin}[x]}{x^2}$

2     $\dfrac{-0.0130396 - x \, \mathrm{Cos}[x] + \mathrm{Sin}[x]}{x^2}$

3     $\dfrac{0.480441 - x \, \mathrm{Cos}[x] + \mathrm{Sin}[x]}{x^2}$

4     $\dfrac{0.973921 - x \, \mathrm{Cos}[x] + \mathrm{Sin}[x]}{x^2}$

5     $\dfrac{1.4674 - x \, \mathrm{Cos}[x] + \mathrm{Sin}[x]}{x^2}$

```
Plot[Evaluate[Table[y[x, 0.2 * j], {j, 1, 5}]], {x, 0.1, 0.2}];
```

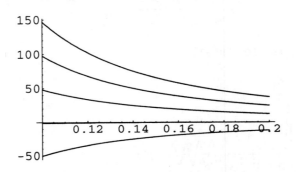

```
Plot[Evaluate[Table[y[x, 0.2 * j], {j, 1, 5}]], {x, 0.2, 10}];
```

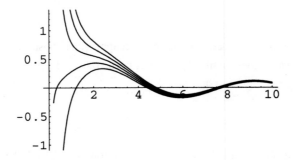

■ **(e)**

The solutions in (d) approach either ∞ or $-\infty$ as x approaches 0 from the right, and approach 0 as x approaches ∞. Can we find a solution that has no singularity at $x = 0$? The general solution is given at the beginning of this problem. The only way that solution could possibly be nonsingular at $x = 0$ is if $c = 4/\pi^2$. We can check using L'Hôpital's rule that in this case the function is continuous at 0. Here is a graph of this solution.

```
Plot[y[x, 4 / Pi^2], {x, -5, 5}];
```

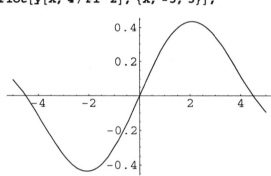

■ **Problem Set B, Problem 4**

We are interested in the solutions to $y' = (x - e^{-x})/(y + e^y)$, specifically the solution satisfying $y(1.5) = 0.5$.

■ **(a)**

```
Clear[y]
DSolve[y'[x] == (x - Exp[-x]) / (y[x] + Exp[y[x]]), y[x], x]
```

Solve::tdep : The equations appear to involve transcendental
    functions of the variables in an essentially non-algebraic way.

$$\text{Solve}\Big[\frac{1}{2} E^{-x} (-2 + 2 E^{x+y[x]} - E^x x^2 + E^x y[x]^2) == C[1], \{y[x]\}\Big]$$

The solutions of the equation are implicit, and the warning message indicates that *Mathematica* cannot find an explicit solution. To use the implicit solution, we define a function $f(x, y)$ equal to the left-hand side of the equation reported by *Mathematica*. The solution of the IVP is obtained by setting this function equal to $f(1.5, 0.5)$.

```
Clear[f, y, c]
f[x_ , y_] =
  (1 / 2) Exp[-x] (-2 + 2 Exp[x + y] - x^2 * Exp[x] + y^2 * Exp[x]) //
  Simplify
```

$$-E^{-x} + E^{y} - \frac{x^2}{2} + \frac{y^2}{2}$$

```
c = f[1.5, 0.5]
```

```
0.425591
```

■ (b)

We can use **ContourPlot** to plot some of the solution curves.

```
ContourPlot[f[x, y], {x, -1, 3}, {y, -2, 2}];
```

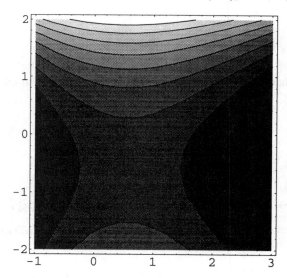

■ (c), (d)

Here are the values of the solution of the IVP at the points $x = 0, 1, 1.8,$ and 2.1.

```
FindRoot[f[0, y] == c, {y, 0.5}]
```

$\{y \to 0.318386\}$

```
FindRoot[f[1, y] == c, {y, 0.5}]
```

$\{y \to 0.235633\}$

```
FindRoot[f[1.8, y] == c, {y, 0.5}]
```

$\{y \to 0.682187\}$

```
FindRoot[f[2.1, y] == c, {y, 0.5}]
```

$\{y \to 0.866212\}$

The following graph shows the contour corresponding to the value $c = f(1.5, 0.5)$. The top curve corresponds to the solution of the IVP because it passes through the point $(1.5, 0.5)$. We mark the points with x-coordinates $0, 1, 1.8,$ and 2.1.

```
<< Graphics`ImplicitPlot`
Show[solcurve, Graphics[
   {PointSize[0.03], Point[{0, 0.32}], Point[{1, 0.24}],
    Point[{1.8, 0.68}], Point[{2.1, 0.87}]}]];
```

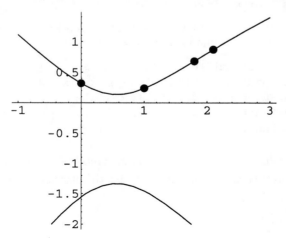

■ Problem Set C, Problem 3

We are going to investigate the IVP $y' = (y - x)(1 - y^3)$, $y(0) = b$, for various choices of the initial value b. We first try to find the exact (symbolic) solution with **DSolve**.

```
Clear[y]
DSolve[{y'[x] == (y[x] - x) * (1 - y[x]^3), y[0] == b}, y[x], x]
```

```
DSolve[{y'[x] == (-x + y[x]) (1 - y[x]³), y[0] == b}, y[x], x]
```

We see that **DSolve** is not able to find the exact solution. We thus solve the problem numerically.

In the next cell the first line clears y; the second line creates a function (with delayed evaluation) representing the solution of the IVP for a given value of the initial condition $y(0) = b$; and the last line makes a list of solutions for $b = 0$, 0.5, 1, 1.5, and 2. We create this list so that *Mathematica* does not have to run **NDSolve** every time we use the function $y(x, b)$.

```
Clear[y]
y[x_, b_] :=
 y[x] /. First[NDSolve[{y'[x] == (y[x] - x) * (1 - y[x]^3),
     y[0] == b}, y[x], {x, 0, 30}]]
sollist = Table[y[x, b], {b, 0, 2, 0.5}]
```

```
NDSolve::ndsz : At x == 1.18795177275153296`, step size
    is effectively zero; singularity suspected.
```

```
NDSolve::ndsz : At x == 2.1196605337876111`, step size
    is effectively zero; singularity suspected.
```

```
{InterpolatingFunction[{{0., 1.18795}}, <>][x],
 InterpolatingFunction[{{0., 2.11966}}, <>][x],
 InterpolatingFunction[{{0., 30.}}, <>][x],
 InterpolatingFunction[{{0., 30.}}, <>][x],
 InterpolatingFunction[{{0., 30.}}, <>][x]}
```

Note that with the first two solutions, **NDSolve** stops before it reaches $x = 30$. As we shall see from the graphs, this is because these solutions have singularities near the indicated values.

▪ (a)

Here is a plot of the five numerical solutions. The lowest curve corresponds to $b = 0$ and the highest to $b = 2$. When we execute this **Plot** command, we get several warning messages because we are trying to plot beyond the intervals of existence of the first two solutions (which are defined up to 1.18... and 2.11..., respectively). We have deleted these warning messages. (*Note*: We are definitely not suggesting that you ignore warning messages. You should look carefully at any warning messages that *Mathematica* supplies and try to understand why they occur.)

```
plot1 =
 Plot[Evaluate[sollist], {x, 0, 4}, PlotRange -> {-3, 4}];
```

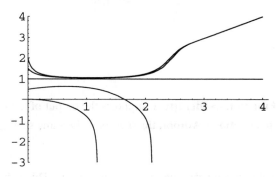

▪ (b)

The plot indicates that if $b = 1$, the corresponding solution is $y = 1$; if $0 < b < 1$, the solution first increases and then decreases to $-\infty$; if $b \le 0$, the solution decreases to $-\infty$; and if $b > 1$, the solution first decreases and then increases to $+\infty$. That $y = 1$ is the solution corresponding to $b = 1$ can be verified by substituting $y = 1$ into the IVP. It also appears that for values of b less than 1, the solutions have vertical asymptotes at finite values of x. One cannot be certain about this from the plots, but it is in fact true.

▪ (c)

Now we'll combine our graphs with the graph of the line $y = x$.

```
plot2 = Plot[x, {x, 0, 4},
    PlotStyle -> Dashing[{0.03}], DisplayFunction -> Identity];
Show[plot1, plot2, AspectRatio -> Automatic];
```

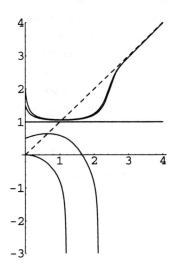

We used the option **DisplayFunction -> Identity** to suppress the output of the first **Plot** command, and the option **AspectRatio -> Automatic** to enforce the same scale on the two axes.

It appears that for $b > 1$, the solutions are asymptotic to the line $y = x$. This is further clarified by plotting the three larger solutions over a longer interval.

```
Plot[Evaluate[sollist[[{3, 4, 5}]]], {x, 0, 30},
  AspectRatio -> Automatic];
```

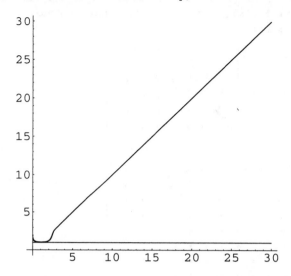

To understand this asymptotic behavior, note that along the line $y = x$, y' is zero; above this line y' is negative; and below the line y' is positive. Thus the solutions are pushed toward the line $y = x$.

Here is a plot of the direction field for the differential equation, which confirms the observations we have made about the solutions. Notice the horizontal arrow at the upper right grid point; this occurs because the grid point (4, 4) is exactly on the line $y = x$.

```
<< Graphics`PlotField`
PlotVectorField[{1, (y - x) * (1 - y^3)}, {x, 0, 4}, {y, -3, 4},
  ScaleFunction -> (1&), Ticks -> None, Frame -> True,
  AspectRatio -> 1];
```

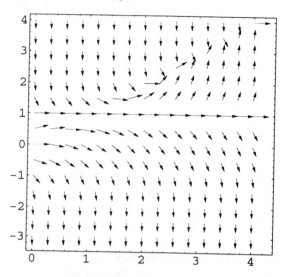

■ Problem Set D, Problem 1

We consider Airy's equation

$$y'' = x\,y$$

with initial conditions $y(0) = 0$, $y'(0) = 1$. We first find the solution.

```
y1[x_] = y[x] /. First[
  DSolve[{y''[x] == x * y[x], y[0] == 0, y'[0] == 1}, y[x], x]]
```

$$-\frac{1}{2}\,3^{1/3}\,\text{AiryAi}[x]\,\text{Gamma}\Big[\frac{1}{3}\Big] + \frac{\text{AiryBi}[x]\,\text{Gamma}\big[\frac{1}{3}\big]}{2\,3^{1/6}}$$

■ (a)

For x close to 0, we want to compare the solution of Airy's equation with the solution to the IVP: $y'' = 0$, $y(0) = 0$, $y'(0) = 1$. The solution of this IVP is $y = x$. Here is a plot of the solution to Airy's equation and the facsimile solution $y = x$. The facsimile solution is plotted in gray.

```
Plot[{y1[x], x}, {x, -2, 2},
  PlotStyle -> {{}, GrayLevel[0.7]}];
```

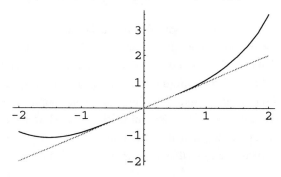

The facsimile solution does indeed agree quite well with the actual solution in a neighborhood of the origin.

■ (b)

Now for x close to $-16 = -(4^2)$, we want to compare the solution of Airy's equation to a facsimile solution of the form $c_1 \sin(4x + c_2)$. We plot the solution of Airy's equation on the interval $(-18, -14)$.

```
Plot[y1[x], {x, -18, -14}];
```

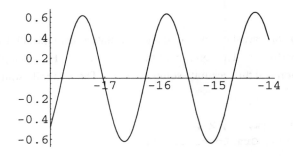

This certainly looks like a sine wave. Let's see how well it matches up with an appropriate sine wave. We have to choose constants c_1 and c_2 in the facsimile solution to make it match up. Note first that c_1 is the amplitude of the facsimile solution, and we can see from the graph that the amplitude of the solution to Airy's equation is about 0.61 on this interval (see the *Graphics* section of Chapter 8 for instruction on finding coordinates of points in *Mathematica* plots). The constant c_2 determines the phase shift and can be read off from the zeros of the solution. In particular, the solution of Airy's equation has a zero at about -16.3, so we should choose c_2 so that $4*(-16.3) + c_2 = 0$; i.e. $c_2 = 65.2$. Here is the plot.

```
Plot[{y1[x], 0.61 Sin[4 x + 65.2]}, {x, -18, -14},
    PlotStyle -> {{}, GrayLevel[0.7]}];
```

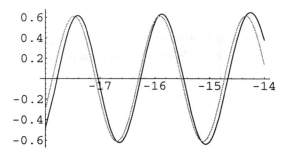

The plots match up very well. Why did we have to choose the values of c_1 and c_2 by hand? Our analysis suggests that the facsimile solution should approximate the actual solution near $x = -K^2$. But the initial condition we've used for Airy's equation is at the point $x = 0$, where the sinusoidal facsimile solutions do not approximate the solution of

Airy's equation. Thus the undetermined constants in the facsimile solution do not really have anything to do with the initial conditions in Airy's equation, so we had to choose them by hand to make the solutions match up. Nevertheless, the *frequency* of the facsimile solution is determined by K and is independent of the undetermined constants. Thus we can conclude at least that the frequency of the solutions of Airy's equation in a neighborhood of $x = -K^2$ will be proportional to K.

■ **(c)**

Now for x close to $16 = (4^2)$, we want to compare the solution of Airy's equation to a facsimile solution of the form $c_1 \sinh(4\, x + c_2)$. As in (b), we first plot the numerical solution of Airy'sequation on the interval (14, 18).

    Plot[y1[x], {x, 14, 18}];

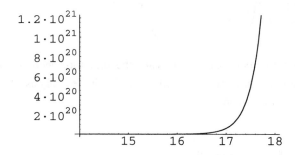

We'd like to compare this graph to that of the hyperbolic sine function. Again, we have to choose c_1 and c_2. Let's make an arbitrary choice this time: $c_1 = 1$, $c_2 = 0$.

    Plot[Sinh[4 x], {x, 14, 18}];

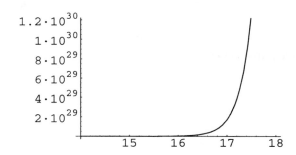

The graphs are remarkably similar. Note, however, that the values in the second graph are about 10^9 greater than in the first, which means we should have chosen c_1 to be about 10^9.

■ **(d)**

```
Plot[y1[x], {x, -20, 2}];
```

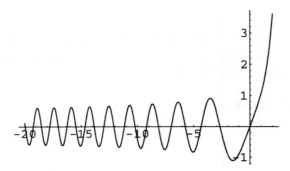

It appears from the graph that the frequency increases and the amplitude decreases as x goes to $-\infty$. The increasing frequency was predicted by our facsimile analysis. The decreasing amplitude is harder to explain.

■ Problem Set E, Problem 1

In this problem, we examine the Taylor approximation of order 10 of the solution of Airy's equation.

■ **(a)**

First we compute the Taylor approximation.

```
Clear[y]
odeseries = Series[y''[x] - x*y[x], {x, 0, 10}]
```

$y''[0] + (-y[0] + y^{(3)}[0]) x +$

$\left(-y'[0] + \dfrac{1}{2} y^{(4)}[0]\right) x^2 + \left(-\dfrac{1}{2} y''[0] + \dfrac{1}{6} y^{(5)}[0]\right) x^3 +$

$\left(-\dfrac{1}{6} y^{(3)}[0] + \dfrac{1}{24} y^{(6)}[0]\right) x^4 + \left(-\dfrac{1}{24} y^{(4)}[0] + \dfrac{1}{120} y^{(7)}[0]\right) x^5 +$

$\left(-\dfrac{1}{120} y^{(5)}[0] + \dfrac{1}{720} y^{(8)}[0]\right) x^6 + \left(-\dfrac{1}{720} y^{(6)}[0] + \dfrac{y^{(9)}[0]}{5040}\right) x^7 +$

$\left(-\dfrac{y^{(7)}[0]}{5040} + \dfrac{y^{(10)}[0]}{40320}\right) x^8 + \left(-\dfrac{y^{(8)}[0]}{40320} + \dfrac{y^{(11)}[0]}{362880}\right) x^9 +$

$\left(-\dfrac{y^{(9)}[0]}{362880} + \dfrac{y^{(12)}[0]}{3628800}\right) x^{10} + O[x]^{11}$

```
coeffs = Solve[{odeseries == 0, y[0] == 1, y'[0] == 0}]
```

$\{\{y[0] \to 1, \ y'[0] \to 0, \ y''[0] \to 0, \ y^{(3)}[0] \to 1, \ y^{(4)}[0] \to 0,$
$\quad y^{(5)}[0] \to 0, \ y^{(6)}[0] \to 4, \ y^{(7)}[0] \to 0, \ y^{(8)}[0] \to 0,$
$\quad y^{(9)}[0] \to 28, \ y^{(10)}[0] \to 0, \ y^{(11)}[0] \to 0, \ y^{(12)}[0] \to 280\}\}$

```
seriessol = Series[y[x], {x, 0, 10}] /. First[coeffs]
```

$1 + \dfrac{x^3}{6} + \dfrac{x^6}{180} + \dfrac{x^9}{12960} + O[x]^{11}$

```
answer = Normal[seriessol]
```

$1 + \dfrac{x^3}{6} + \dfrac{x^6}{180} + \dfrac{x^9}{12960}$

■ (b)

Now we compute the exact solution to the initial value problem using **DSolve**.

```
Clear[w]
airy = DSolve[{w''[x] - x*w[x] == 0, w[0] == 1, w'[0] == 0},
  w[x], x]
```

$\left\{\left\{w[x] \to \right.\right.$
$\quad \left.\left. \dfrac{1}{2} 3^{2/3} \ \text{AiryAi}[x] \ \text{Gamma}\left[\dfrac{2}{3}\right] + \dfrac{1}{2} 3^{1/6} \ \text{AiryBi}[x] \ \text{Gamma}\left[\dfrac{2}{3}\right]\right\}\right\}$

```
a[x_] = w[x] /. First[airy]
```

$$\frac{1}{2} \ 3^{2/3} \ \text{AiryAi}[x] \ \text{Gamma}\left[\frac{2}{3}\right] + \frac{1}{2} \ 3^{1/6} \ \text{AiryBi}[x] \ \text{Gamma}\left[\frac{2}{3}\right]$$

The solution is given in terms of the built-in functions **AiryAi** and **AiryBi**, which form a fundamental set of solutions of Airy's equation. The solution also includes constants expressed in terms of the built-in **Gamma** function. (We could convert these constant to decimal numbers by using **N**.)

Now we compare the Taylor polynomial approximation to the exact solution produced by **DSolve**.

```
Plot[{answer, a[x]}, {x, -10, 0},
  PlotStyle -> {GrayLevel[0.7], {}}];
```

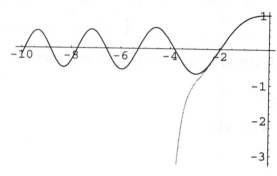

```
Plot[{answer, a[x]}, {x, 0, 5},
  PlotStyle -> {GrayLevel[0.7], {}}];
```

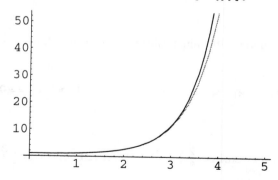

By plotting the ratio of the approximate and exact solutions, we can get a good picture of the relative accuracy of the approximation.

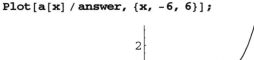

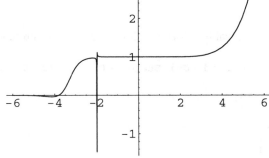

When the ratio is close to one, the approximation is relatively accurate (the vertical line in the graph appears because the Taylor polynomial has a zero near -2). Thus the Taylor approximation is accurate roughly on the interval $(-2, 3)$.

■ **(c)**

The general solution of the equation $y'' = y$ is $y(x) = A\,e^x + B\,e^x$, which grows exponentially if A is not zero. Thus a generic solution of Airy's equation grows exponentially for $x > 0$. Since a polynomial grows much more slowly than an exponential, we should expect that a polynomial approximation of the solution to Airy's equation will diverge after a short time from the actual solution. This is what we observed in part (b).

■ Problem Set E, Problem 16

Here is the program from Chapter 11 for computing solutions of initial value problems using the Laplace Transform method.

```
<< Calculus`LaplaceTransform`
LTSolve[eqn_, y0_, yp0_] := Module[
   {lteqn, ltsol, Y, s}, lteqn = LaplaceTransform[eqn, t, s] /.
     LaplaceTransform[y[t], t, s] -> Y[s];
   ltsol = Solve[{lteqn, y0, yp0}, Y[s]];
   InverseLaplaceTransform[Y[s] /. First[ltsol], s, t]]
```

■ (a)

We define a square wave on the interval $[0, 10\pi]$ by using a sum of unit step functions.

```
h[t_] = -1 + 2 * Sum[(-1)^i * UnitStep[t - Pi * i], {i, 0, 10}];

Plot[h[t], {t, 0, 30}];
```

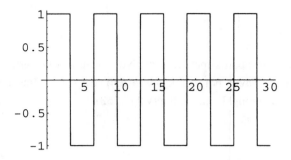

■ (b)

```
ode = y''[t] + y[t] == h[t]
```

y[t] + y″[t] ==
  -1 + 2 (UnitStep[t] + UnitStep[-10 π + t] - UnitStep[-9 π + t] +
     UnitStep[-8 π + t] - UnitStep[-7 π + t] + UnitStep[-6 π + t] -
     UnitStep[-5 π + t] + UnitStep[-4 π + t] - UnitStep[-3 π + t] +
     UnitStep[-2 π + t] - UnitStep[-π + t])

```
solb[t_] = LTSolve[ode, y[0] == 0, y'[0] == 1]
```

1 - Cos[t] + Sin[t] + 2 (1 - Cos[t]) UnitStep[-10 π + t] -
  2 (1 + Cos[t]) UnitStep[-9 π + t] +
  2 (1 - Cos[t]) UnitStep[-8 π + t] -
  2 (1 + Cos[t]) UnitStep[-7 π + t] +
  2 (1 - Cos[t]) UnitStep[-6 π + t] -
  2 (1 + Cos[t]) UnitStep[-5 π + t] +
  2 (1 - Cos[t]) UnitStep[-4 π + t] -
  2 (1 + Cos[t]) UnitStep[-3 π + t] +
  2 (1 - Cos[t]) UnitStep[-2 π + t] -
  2 (1 + Cos[t]) UnitStep[-π + t]

```
Plot[{solb[t], h[t]}, {t, 0, 30}];
```

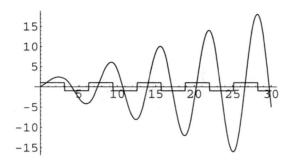

We do see resonance. The solution appears to grow without bound. Let's check the solution by using **NDSolve.**

```
numsolb[t_] = y[t] /. First[
    NDSolve[{ode, y[0] == 0, y'[0] == 1}, y[t], {t, 0, 30}]]
```

```
InterpolatingFunction[{{0., 30.}}, <>][t]
```

```
Plot[numsolb[t], {t, 0, 30}];
```

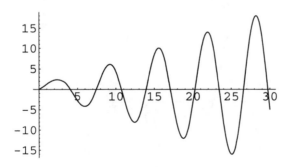

■ **(c)**

```
Off[General::spell1]
solc[t_] =
  LTSolve[y''[t] + y[t] == h[t / 2], y[0] == 0, y'[0] == 1];
```

```
Plot[{solc[t], h[t / 2]}, {t, 0, 30}];
```

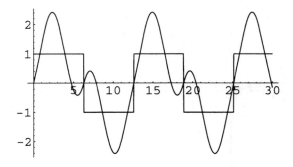

This solution does not exhibit resonance; it remains bounded. We can check the solution using **NDSolve.**

```
numsolc[t_] = y[t] /. First[
    NDSolve[{y''[t] + y[t] == h[t / 2], y[0] == 0, y'[0] == 1},
      y[t], {t, 0, 30}]]
```

```
InterpolatingFunction[{{0., 30.}}, <>][t]
```

```
Plot[numsolc[t], {t, 0, 30}];
```

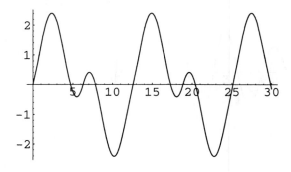

■ **(d)**

Now we examine the solution with forcing function $h(2\,t)$, which has period π.

```
sold[t_] =
  LTSolve[y''[t] + y[t] == h[2 t], y[0] == 0, y'[0] == 1];
```

```
Plot[{sold[t], h[2 t]}, {t, 0, 15}];
```

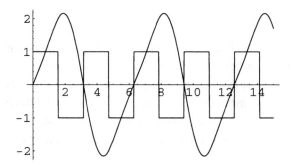

This solution does not exhibit resonance. We can check the solution using **NDSolve**.

```
numsold[t_] = y[t] /. First[
   NDSolve[{y''[t] + y[t] == h[2 t], y[0] == 0, y'[0] == 1},
   y[t], {t, 0, 15}]]
```

```
InterpolatingFunction[{{0., 15.}}, <>][t]
```

```
Plot[numsold[t], {t, 0, 15}];
```

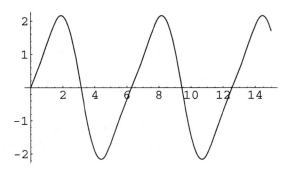

■ **(e)**

The results of parts (b), (c), and (d) indicate that resonance may be a consequence of the period of the forcing function, regardless of the actual shape of the function. This is, in fact, true. A rigorous justification of this fact involves Fourier analysis, which shows that any periodic function can be "built up" out of cosine and sine functions of different periods. Most (but not all!) functions with minimal period 2π contain $\sin(x)$ or $\cos(x)$ as a building block, and it is the occurrence of the $\sin(x)$ and $\cos(x)$ terms that leads to the resonance observed in part (b).

■ Problem Set F, Problem 6

■ (a)

First we find the critical points of the system.

```
Clear[x, y]
crit =
  Solve[{x (1.5 - x - 0.5 y) == 0, y (2 - 1.5 x - 0.5 y) == 0}, {x, y}]

{{x → 0., y → 0.}, {x → 0., y → 4.}, {x → 1., y → 1.},
 {x → 1.5, y → 0.}}
```

We want to analyze the critical points of the system. To do this, we compute the matrix of partials of the system, and then evaluate this matrix and find its eigenvalues at the critical points.

```
sys1 = x (1.5 - x - 0.5 y);
sys2 = y (2 - 1.5 x - 0.5 y);
A = {{D[sys1, x], D[sys1, y]}, {D[sys2, x], D[sys2, y]}}

{{1.5 - 2 x - 0.5 y, -0.5 x}, {-1.5 y, 2 - 1.5 x - 1. y}}
```

```
Eigensystem[A /. crit[[1]]]

{{2., 1.5}, {{0., 1.}, {1., 0.}}}
```

Eigensystem[A /. crit[[2]]]

{{-2., -0.5}, {{0., 1.}, {0.242536, -0.970143}}}

Eigensystem[A /. crit[[3]]]

{{-1.65139, 0.151388},
 {{-0.608894, -0.793252}, {0.398322, -0.917246}}}

Eigensystem[A /. crit[[4]]]

{{-1.5, -0.25}, {{1., 0.}, {-0.514496, 0.857493}}}

By inspecting the eigenvalues at the four critical points, we can classify the critical points as follows:

(0, 0)	unstable node
(0, 4)	asymptotically stable node
(1, 1)	unstable saddle point
(1.5, 0)	asymptotically stable node

■ **(b)**

Here is a plot of the vector field of the system.

```
<< Graphics`PlotField`
csvf = PlotVectorField[{x (1.5 - x - 0.5 y), y (2 - 1.5 x - 0.5 y)},
  {x, 0, 3}, {y, 0, 5}, ScaleFunction -> (# + 50&),
  Frame -> True, AspectRatio -> 1];
```

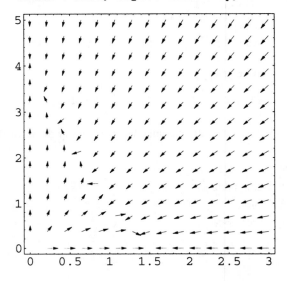

The vector field confirms what we found in part (a).

■ **(c)**

To plot the phase portrait of the system, we use the approach described at the end of Chapter 11.   We use a rectangular grid of 42 different initial conditions.

```
Clear[t, x0, y0]
cspop[tval_, x0_, y0_] := ({x[t], y[t]} /.
    First[NDSolve[{x'[t] == x[t] (1.5 - x[t] - 0.5 y[t]),
        y'[t] == y[t] (2 - 1.5 x[t] - 0.5 y[t]),
        x[0] == x0, y[0] == y0},
        {x[t], y[t]}, {t, 0, 20}]]) /.
  t -> tval
cs[t_] =
 Flatten[Table[cspop[t, i, j], {i, 0, 3, 0.5}, {j, 0, 5}], 1];
ParametricPlot[
 Evaluate[cs[t]], {t, 0, 20}, PlotRange -> {{0, 2}, {0, 5}}];
```

This graph is not ideal. The problem is that a rectangular grid of initial conditions is not the best choice for this system. We can see from the graph that a better choice of initial conditions would be a collection of evenly spaced points along lines parallel to and above and below the line from (0, 4) to (1.5, 0). Since that line has slope about −2.5, we redefine the function **cs** as follows:

```
Clear[cs]
cs[t_] = Flatten[Table[cspop[t, j * x0, j (2.5 - 2.5 x0)],
    {x0, 0, 1, 0.1}, {j, 0.6, 2.6, 2}], 1];
csplot = ParametricPlot[Evaluate[cs[t]], {t, 0, 20},
  PlotRange -> {{0, 2}, {0, 5}}];
```

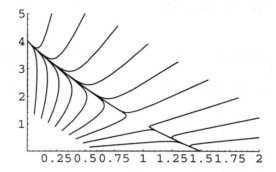

Much better! By putting the parametric plot and the vector field plot together we can see the direction of the trajectories.

```
Show[csplot, csvf];
```

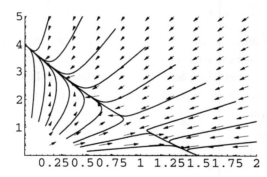

■ (d)

Here is a table of values of the populations. We use the option **TableDepth -> 2** to prevent the **TableForm** command from stripping off all levels of braces.

```
TableForm[Table[Evaluate[{t, cspop[t, 2.5, 2]}], {t, 1, 20}],
 TableHeadings -> {None, {"Time", "Populations"}},
 TableDepth -> 2]
```

Time	Populations
1	{1.29443, 0.761069}
2	{1.20257, 0.629452}
3	{1.20352, 0.569923}
4	{1.2221, 0.520469}
5	{1.24499, 0.471821}
6	{1.2692, 0.422989}
7	{1.29376, 0.374567}
8	{1.31806, 0.327472}
9	{1.34158, 0.282627}
10	{1.36383, 0.240847}
11	{1.38442, 0.202758}
12	{1.40307, 0.168753}
13	{1.41961, 0.13899}
14	{1.434, 0.113408}
15	{1.4463, 0.0917785}
16	{1.45665, 0.0737513}
17	{1.46524, 0.0589121}
18	{1.47228, 0.0468243}
19	{1.478, 0.0370647}
20	{1.4826, 0.0292416}

■ **(e)**

There is no peaceful coexistence because almost all the trajectories tend toward an equilibrium point representing a positive population of one species and a zero population of the other. For example, in (d) we saw that the population of x tends to 1.5 and the population of y tends to 0. The only nonzero initial population distribution which results in no change is $x = 1$, $y = 1$.

To approximate the separatrices, we solve the system for negative time for initial conditions near the unstable equilibrium point (1, 1). From the portrait in (c), we expect one separatrix running from the origin to (1, 1); and another running from (1, 1) out towards infinity. To find the first we choose an initial data point just to the left and below (1, 1) and solve backwards in time; and to find the second we do the same thing with an initial data point above and to the right of (1, 1).

```
Clear[cspop, cs];
cspop[tval_, x0_, y0_] := ({x[t], y[t]} /.
    First[NDSolve[{x'[t] == x[t] (1.5 - x[t] - 0.5 y[t]),
        y'[t] == y[t] (2 - 1.5 x[t] - 0.5 y[t]),
        x[0] == x0, y[0] == y0},
        {x[t], y[t]}, {t, -10, 20}]]) /.
    t -> tval
```

The following command plots the separatrix but generates a lot of warning messages
regarding values of *t* outside the domain of the **InterpolatingFunction** generated by
NDSolve. We turn some of the warning messages off to reduce clutter.

```
Off[InterpolatingFunction::dmval]
separatrixplot = ParametricPlot[Evaluate[
    {cspop[t, 0.98, 0.98], cspop[t, 1.02, 1.02]}], {t, -10, 0},
    PlotRange -> {{0, 2}, {0, 5}}, PlotStyle -> GrayLevel[0.7]];
```

```
NDSolve::ndsz : At t == -2.367, step size is effectively
    zero; singularity suspected.
```

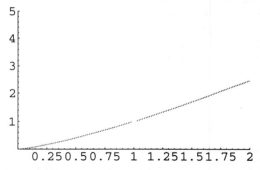

```
Show[csplot, separatrixplot];
```

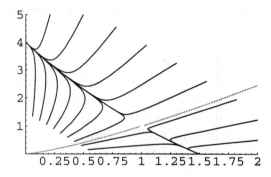

■ **(f)**

To find a numerical approximation to where the vertical line $x = 1.5$ cuts the separatrix
we delineated above, we first observe that the value is between 1.5 and 2.0. So we plot
trajectories beginning at $x = 1.5$ and $y = 1.5, 1.6, ..., 2.0$. Each trajectory produces a
warning message like the one above from **NDSolve**, indicating that a singularity is
encountered for negative t. We use the **Off** command again to suppress these messages.

```
Off[NDSolve::ndsz]
Clear[cs]
cs[t_] = Table[cspop[t, 1.5, y0], {y0, 1.5, 2.0, 0.1}];
ParametricPlot[Evaluate[cs[t]], {t, 0, 10},
  PlotRange -> {{0.5, 2}, {0.5, 2}}];
```

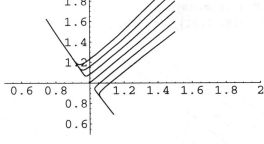

It appears that the desired y value is very close to 1.7. We now refine the picture using
the same strategy. We use the **PlotRange** option to clarify the portrait.

```
cs[t_] = Table[cspop[t, 1.5, y0], {y0, 1.68, 1.72, 0.01}];
ParametricPlot[Evaluate[cs[t]], {t, 0, 10},
 PlotRange -> {{0.95, 1.05}, {0.95, 1.05}}];
```

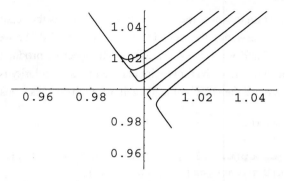

The desired value is between 1.69 and 1.70. One more iteration will settle the matter!

```
cs[t_] = Table[cspop[t, 1.5, y0], {y0, 1.69, 1.70, 0.002}];
ParametricPlot[Evaluate[cs[t]], {t, 0, 10},
 PlotRange -> {{0.995, 1.01}, {0.995, 1.01}}];
```

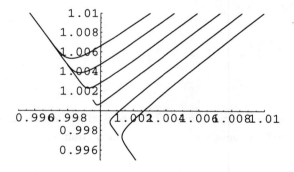

Voila. The answer is 1.69.

We could, of course, continue this process and obtain greater and greater accuracy, but it is not possible to find the value exactly. Even if we were given the exact value it would not be possible to verify its correctness. This is because **NDSolve** is an approximate method and errors inevitably occur—errors that cause the trajectory to fall off the separatrix, and then to tend toward one of the stable equilibria.

Index

The index uses the same conventions for fonts that are used throughout the book. *Mathematica* commands, such as **DSolve**, are printed in boldface. Menu options, such as `File`, are printed in a monospaced typewriter font. Trademarked names, such as *Windows 95*, are printed in a slanted font. Everything else is printed in a standard font.